RETURNING TO NOTHING

For S. K. Jeffery

RETURNING TO NOTHING

THE MEANING OF LOST PLACES

Peter Read

Department of History
Australian National University

CAMBRIDGE UNIVERSITY PRESS
Cambridge, New York, Melbourne, Madrid, Cape Town, Singapore, São Paulo, Delhi

Cambridge University Press
The Edinburgh Building, Cambridge CB2 8RU, UK

Published in the United States of America by Cambridge University Press, New York

www.cambridge.org
Information on this title: www.cambridge.org/9780521576994

First published 1996
Re-issued in this digitally printed version 2009

A catalogue record for this publication is available from the British Library

National Library of Australia Cataloguing in Publication data

Read, Peter, 1945– .
Returning to nothing: the meaning of lost places.
Bibliography.
Includes index.
1. Extinct cities – Psychological aspects. 2. Environmental
psychology. I. Title.
155.9

Library of Congress Cataloguing in Publication data

Read, Peter, 1945– .
Returning to nothing: the meaning of lost places/Peter Read.
p. cm.
Includes bibliographical references and index.
1. Natural disasters – Australia – Psychological aspects.
2. Landscape changes – Australia – Psychological aspects. 3. Loss
(Psychology). 4. Grief – Australia. I. Title.
GB5011.89.R43 1996
155.9′35–dc20 96–16590

ISBN 978-0-521-57154-8 hardback
ISBN 978-0-521-57699-4 paperback

CONTENTS

Liubov Goodbye dear house, old grandfather house. Winter will pass, spring will come again, and then you won't be here any more, you'll be pulled down. How much these walls have seen!

A. Chekhov, *The Cherry Orchard*

'And the bulldozers are here, I see?'
The neighbours went
One by one,
The Johnson's house next door:
One day, men on the roof
Throwing down tiles
Smashing them in the garden.
Curtains still at the windows
Carpet on the floor
All ruined.
Bulldozer in the living room
pushing down walls.
Ken loved his pool
Cleaned it every day.
The bulldozer dug it up in an afternoon.

Julia Garnett, Extract from 'The Turpentine Tree'
Orchard Rd, Beecroft
February 1995

PREFACE

The places round the billabong
are pretty much the way they were
but like a lot of things, they're gone.
Philip Hodgins[1]

Hodgins wrote these words after he returned to the locations of his childhood to find them unrecognisable, changed or destroyed. This book is a history of the migrations away from dying homes, streets, neighbourhoods, suburbs, towns, cities and countries—and the return journeys to the empty spaces where once they were. I have called these journeys, which can take place either on the ground or in the mind, 'returning to nothing'.

Often the journey to nothing is actual. Kass Hancock returned to Darwin twenty years after Cyclone Tracy had destroyed her home and almost the entire suburb of Wagaman. The house opposite the place she had lived, where two little children had been killed, had vanished and the street itself was almost unrecognisable. Cracks in the pavement in Wagaman Terrace were almost the only tangible reminder of many sunny days and of one dreadful night.[2] The returning journey to a lost place can also be metaphorical. Dorothy Hewett, in her poem 'In Summer', imagined herself going back for a look to her childhood home, listening to the ever-flowing rip in the darkness outside as it closed about the family and the dwelling. She saw herself walking among her brothers and sisters playing 'Ricketty Kate', telling them which card to play—with the exact knowledge of what unpleasant fates would befall each of her family members.[3]

Returning to Nothing owes its immediate origins to a project, sponsored by the Australian Research Council, into the ways in which Australians relate to and value country of significance to them. I had already worked with several rural families for more than a year studying contemporary attachments, when, while holidaying near the site of the drowned town of Old Adaminaby in the southern highlands of New South Wales, I bought a book of local history entitled *It Doesn't Snow like it used to*.[4] In the first chapter was part of a poem called 'Adaminaby The Old Town'.

> The stations, farms, the sheds, the barns,
> Are all awash in the deep waters
> While ghosts of men swim through the glen,
> Drowned faces haunt their sons and daughters.[5]

The theme of this book was decided at that moment. Half a dozen trips to talk to the mourning former residents of Adaminaby confirmed what Douglas Stewart meant by:

> the mystery and the pathos
> that seep from earth and bubble out from water
> In any place where men have lived and bred
> And feuded with each other.[6]

I had embarked on a journey which to most of the people who spoke to me was a painful process of mourning and grief, and to me was also one of enlightenment. Almost always accompanied by someone to whom the site was dear, I explored the ruins or sites of homesteads and country towns, houses which had been sold and the locations of former parks and neighbourhoods. For the last chapter I began work in Beecroft, Sydney, where several home owners whose suburb lay in the path of the F2 Freeway shared their feelings of their last and hopeless battle against the New South Wales Roads and Traffic Authority.

My own explorations of lost places began when, as a fourteen-year-old growing up on the north shore of Sydney, I was fascinated by stories of two of my ancestors. Robert and Selina Pockley had built a house called Lorne in Killara, which in 1924 had been compulsorily purchased and demolished for the construction of a reservoir. Though this had occurred more than forty years previously, I felt that its destruction continued to haunt my older relatives. My grandmother knew which article of furniture had gone to which relative and was able to draw an exact plan of the house. My mother had a collection of press clippings about the estate and its controversial resumption, which I stuck into a scrapbook and wrote about. I made my first journey to nothing: I found no trace of Lorne besides the avenue of that name which runs beside the reservoir. Lorne began to haunt me, though in a much less personal way than it did those to whom the house was a living memory. My mother gave me a painting of an enormous azalea bush executed the year before the estate was resumed. Had it been painted because the artist knew that soon it must be destroyed?

My granny revealed that prior to building Lorne, the Pockleys had lived in a house called Doohat in North Sydney. Now I rushed to that dead place also. Like Lorne, its presence was commemorated by a named avenue. Unlike Lorne, the site was still unoccupied even in the busy Miller Street of 1959. Half a dozen pine trees towered about the foundations of what had been a substantial home. Heart thumping, I explored the site, and peeping over a back fence found part of a weathered sandstone fountain which obviously didn't belong to the small suburban garden in which it now stood. Click went the Box Brownie, and another artefact was added to the breathless historian's mental cupboard of lost places. Thirty-six years later my heart was thumping again, as camera in hand I surveyed the wreckage of the drowned town of Adaminaby, unexpectedly and harshly revealed by the drought which had drained Lake Eucumbene to 25 m below high water mark. Who, in the previous week or two, had gathered all those pieces of broken crockery and arranged them on top of a foundation stone? And why? A thumping heart notwithstanding, these wrecked and dead places were other people's grief, not mine. Sadness over a destroyed building of no immediate personal connection was not the savage force which has drawn so many others back on more personal journeys to nothing.

The contributors in a sense chose themselves. Every conversation or interview seemed to conclude with a suggestion that I should visit a friend who had suffered a similar bereavement in another town, suburb or country. Each demolished home I stood before could have been one of thousands, each town site could have been one of hundreds, each destroyed suburb could have been one of dozens. Most countries in the world are lands of forced exile to somebody now living in Australia. My study of Darwin after the cyclone could have been a study of Newcastle after the earthquake. It should not reduce the intensity of the emotions quivering in this book to reflect that they are paralleled in the minds of probably hundreds of thousands of Australians.

The emotions expressed to me were raw, and so is the book. Except in the case of the inundation of Lake Pedder, I have not attempted to set such recollections into the current context of memory or place theory. Possibly these many accounts will be the quarry of further theorisation of oral history and belonging; they may widen our current understanding of the phenomenon of grief and loss. Indeed many questions proceed from these stories. Is there a maximum level of grief which the human psyche allows itself to sustain? Can an ethical analogy be drawn between a bushfire victim's loss of home and a Holocaust survivor's loss of extended family, birthland and the national culture? Are people who own their homes likely to love them more than those who rent them? Is class difference relevant to place studies? Is forced evacuation a necessary condition of grief for a lost place? Do residents of farms love their places more than residents of suburbs? Do children remember places differently from adults? How and why do place memories change at different phases of life? Are women likely to mourn lost places more readily than men? While there are many implied responses in the text, there are no definitive answers. I have left the hundred or so people whose stories are represented in this book largely to speak their own intuitive truths in the poetic and open-ended terms in which they were rendered to me. With place and memory theory I have no quarrel, and have often entered the debate myself in other contexts.[7] But not here, not among such raw and intimate memories, not about these dead places. If this book demonstrates the complexity and depth of feelings for lost places in Australia, its mission will be accomplished.

There is another looming theme about which I have also added little. Each of the individuals and groups represented here have been left to mourn their lost places alone. While developers may have wanted to destroy a town or neighbourhood, they were not emotionally attached to those places and generally do not now contest their memorialising. Those who have embarked on the long journey to nothing at least found nobody standing at the same dead site to challenge the legitimacy of their grief.

In modern Australia, and still more elsewhere in the world, such uncontested attachment is becoming a luxury. On the left bank of the Jordan, in South Africa, Bosnia, Hungary, Uganda, Hong Kong and Cuba, dozens of cultural, ethnic and familial rivals, mostly deprived of their places of attachment, dispute not only physical possession but others' rights of emotional attachments to the same place. We know well that Australians dispute the expansion of suburb over farmland and new houses over old, but we don't so know so well—because such matters are private—that people nurse their own personal and different memories, attachments and griefs. Occasionally these shared but conflicting emotional attachments become public.

In this book I have confined my discussion of emotionally contested places to Namadgi National Park in the Australian Capital Territory. At Namadgi, environmentalists and former pastoralists do not dispute the right to be there, for the environmentalists are firmly in control; the pastoralists seek the right to be remembered and celebrated at the exact location of their achievements. Related to this deepening theme, of course, are separate Aboriginal and non-Aboriginal emotional and spiritual attachments to the same places. Some New Zealand farmers have argued before the Waitangi Tribunal that they, not the Maori claimants, are the true inheritors of the high country, for they have loved it and cherished it for nearly 200 years. Australian farmers are beginning to advance their own sets of valued memories, attachments and histories over the same areas claimed by Aboriginal people. Having worked for many years among Aborigines deprived of their country, and more recently with non-Aborigines deprived of theirs, I am filled with anxiety at the complexity of such disputed attachments. They await a second study, which will follow this book.

For the rip, as Dorothy Hewett put it, will not stop flowing. Our relationships to our own lost places grow more insistent, yet in many Asian cities no building is allowed to remain standing for more than perhaps twelve years. Famous Australian buildings can disappear almost overnight. Whole suburbs vanish before freeway constructions in a matter of weeks.

Philip Hodgins' poem ends:

> I used to walk along
> the rabbit tracks and check my snares
> at places round the billabong.

> But now they keep the rabbits down
> with bait. It's hard to come this far
> and find a lot of things have gone,
>
> that all the snares have been undone
> and what I wanted isn't here.
> These places round the billabong
> are like a lot of things, they're gone.

Each of us has our own remembered billabong which is not the same as once it was. The emotions which emerge from these stories are frightening in their intensity. Let us not underestimate the effect which the loss of dead and dying places has on our own self-identity, mental well-being and sense of belonging.

Peter Read

ACKNOWLEDGMENTS

I am grateful to the the Australian Research Council, which made this work possible by the award of a Fellowship and travel funds to visit the many communities and individuals who took part in this work. The Department of History at the Australian National University provided a friendly and helpful working environment and my colleagues took part in many stimulating discussions and suggestions.

I thank the many individuals who by talking or journeying revisited their lost places of significance. Without them this book would not have been possible. In particular I thank Esme Blackmore, Julia Garnett and Ivan and Jenny Lewis for working with me at particularly difficult times of their lives, Sally Roberts for drawing on the excellent resources of the ANU's North Australian Research Unit, Debbie Rose for her offer of a NARU Visiting Fellowship, Meredith Fletcher for expert knowledge and organisation of the resources of the Centre for Gippsland Studies, and Mark Cranfield of the National Library's Oral History Unit for equipment and other help. For accommodation or meals as well as willingness over many months to share their perceptions of the past, I thank Nita Stewart, Prue and Val McGoldrick, Peter and Betty Boekel, Peter and Sue Boekel, Granville and Rae Crawford, Marge and Keith Mackay, Margaret and Jim Johnson, Paddy and Jan Kerrigan. Some of the people who spoke with me do not appear in this book but their contribution to my understanding of losing a loved place nonetheless has been great. Tim Bonyhady, Sylvia Deutsch, Paula

Hamilton, Ken Taylor and Marivic Wyndham read some or all of the text, and their contributions and suggestions were invaluable. Jay Arthur, as always, felt as well as thought her way through the manuscript. Her contribution to the meaning of this book is never far away.

CHAPTER 1 | LOSING WINDERMERE STATION

SINKING ROOTS

Margaret Johnson spent most of her youth at Narrandera in south-western New South Wales. From her birth in 1933 she was a solitary child. When she thinks of that time she recalls a sleepy town, heat and dust, birds, long summer days, walking by herself knee-deep through piles of plane-tree leaves.[1]

As a young woman Margaret Johnson travelled overseas, and on her return felt restless. Office work held no attraction. At the age of twenty-two she took a job as governess on a sheep property near Young on the central slopes and plains of New South Wales. Here was that half-forgotten but familiar *lovely smell of hot weather and dusty roads*. The following year she married Jim Johnson, the owner of a neighbouring property. The young bride saw the property's overgrown tennis court, dilapidated gravel paths, run-down gardens, horses grazing a few metres from the back door, the dark, two-storey rambling house with heavy curtains and brown blinds. She fell in love with them all, she says, almost instantly.

The property she had come to was 2500 ha of pleasant grazing country. Its name was Windermere Station; it had been held by the Johnson family since 1923. In the nineteenth century ten people had worked on it, now there were two. Except for the river flats the land was, and is, undulating to rough. Margaret Johnson thought it *really quite beautiful because it's got the hills almost on three sides and a lovely creek running down through the middle and little creeks running down into that. Quite lovely.* Out of the south west rolled the wondrous clouds. Above the flats they were *whimsical and friendly*, on the hills they were *intimate and misty*, and when they roared over the Illuni Range they were *quite stupendous and magnificent*. In the paddocks the sky, the air and the soil were part of *the real physical tie to the land, a feeling that is part of your spirit that's divorced from all arguments of logic and reason and behaviour.* Margaret Johnson felt herself in love with *the feeling of the place, that country feeling, that feeling that you are able to exist in a place with not many people; you could call it lonely but you're never lonely. I suppose it's the spirit of the place. It's a very calming sort of feeling to be out amongst those lovely gum trees and the creek and the birds. I've always felt an affinity with that sort of place*. The days were full and rich *without you having to do very special things*. The first familiarities of Windermere were the contours and creeks; attachments deepened year by year to bonds with all the property and the special places within it. Over the next four decades, Margaret Johnson *did a fair job of knowing every gully intimately*.

How do humans form such powerful and mysterious attachments to country? The philosopher Gaston Bachelard believed that all really inhabited space bears the essence of the nature of home, that the human imagination begins to create a recognisable place wherever people find the slightest shelter, walls of impalpable shadows or the illusion of protection.[2] Humans, and apparently other mobile beings, are able and feel the necessity to turn space into place, to identify a site as in some way different from other sites, to erect mental boundaries around it, to live or work in it, or call it home. One conclusion which emerges from the enormous contemporary literature of place studies is that the ways in which humans demarcate their space are bound by the rules and customs of the cultures of which they form a part—the way in which they actually and symbolically create landscape within the cultural community probably reflects other organising principles of that society and its world view.[3] In this book I take it that anything that individuals recognise as 'a place' has been in part constructed to suit them and in part has been created by wider issues of power, group dynamics, conflicting ideologies and institutions. This shaping of identifiable sites affects both the physical appearance of places (for example, boundary fences and street signs), and the way they are conceptualised (such as 'home', 'Melbourne' or 'the bush').[4]

Who forms attachment to place? Individual variations seem as great as different cultural expressions. To the descendants of the invaders, the sense of belonging takes many forms:

> Some of us feel at home nowhere
> Others in one generation fuse with the land.[5]

Aborigines maintain a mutually sustaining relationship:

> We belong to the land; our birth does not sever the cord of life which comes from the land.[6]

Other cultures also take rootedness for granted. The ancient Romans conceived a powerful relationship between gods and the soil. A family erected its own focus, or sacred hearth fire, from which the family took root. As long as the family lasted, the household god expressed in the sacred fire was thought to possess the soil. In this conjoint relationship, both soil and family were immovable, connected to each other through the mystic fire.[7] Other cultures allow attachments to place to form and be expressed collectively and unself-consciously. The Irish have been preoccupied with the nature of place since ancient times. 'They lived off, moved across it, above all named it, with that associative and magical potency which wove it intimately into their lives.'[8] The English may be thought by the Irish to be rational, scientific, sceptical, too out of tune with the earth's vibrations mystically to bond with it—but, though collectively the love of place may form a less dominant theme in English culture, there may be many thousands of English individuals who love the English landscape but who do not find it easy to express their emotions as passionately as the Irish express theirs.

People respond individually to locality, then, and the culture with which they are familiar helps to enlarge, diminish, shape or transform it. Senses of belonging are allied to attachment and love, but the country must first become known and apprehended; as Paul Carter puts it, nature had first to be conceptualised as a place before it could be loved.[9] Cartographers inscribe maps as if the geographic features the Australian explorers stumbled upon were already there, as if rivers flowed waiting for a European to name them. But to recognise a river as a discrete, nameable geographical entity is a cultural, not a natural, expression. There is no such thing as a 'river' until we recognise it as such and place it in the named and identified category of 'river'.

Once such entities are recognised within the meanings of the identifying culture, humans proceed to utilise the familiar geographical features for their individual purposes. Australian Christians erect churches on hills and call the ground sacred, Italian Australians declare certain springs to be holy wells, states declare their boundaries to begin at rivers. Universally recognised entities like the sun take on local cultural significance: Mena Abdullah's Punjabi father, farming on the Gwydir River in New South Wales in the 1950s, used to tell his children that the flaming sunset over the New England tableland was the glory of Allah.[10] Holy wells and an Australian Allah may seem alien to Australians of Anglo-Saxon descent, but these cultural expressions of landscape, though different from the more familiar cultural markers of pastoral property, fence, suburb and cleared land, spring from a similar human impulse. Different human cultures recognise different features in the landscape and imbue them with different cultural characteristics.

After the Australian explorers there came the generation of landholders who erected boundary fences to mark their space and territory: enclosure was essential to Europeans, Carter believed, not only to the act of settling but also to the description of settling.[11] The enclosing fence was followed by the clearing of the home paddock which, like the marked boundaries, seemed as important a symbolic act of possession as a physical necessity. John Dunmore Lang, the famous nineteenth-century Presbyterian minister, voiced the feeling of many Australians of last century:

> Yet all is wild and waste, save where the hand
> Of man, with long continued toil and care,
> Has won a little spot of blooming land
> From the vast cheerless forest here and there.[12]

Clearing meant symbolic possession, hard labour and vivid memories transmitted for generations. Mary Fullerton of Gippsland recalled:

> That was the sort of land that had to be cleared and prepared for the plough by the indomitable settlers along that whole river valley ...

> [Nature] certainly had clad those flats well with tree and shrub and fern. Many an age had she laid down her sowings by the tillage scheme of gravitation, of flood waters, and all those other wonderful and inexhaustible powers of hers whereby the earth is made ready for man. And then man, to do his own sowings, had to remove Nature's, necessity stopping at no desecration, no waste. We had the privilege of seeing ... of helping in the undressing of the loam, and the taking from it of man's first fruits.[13]

In this way non-Aboriginal Australians began to sink their roots of attachment into the Australian earth. It seems something of a myth that non-indigenous Australians did not love the landscape they found and made because it was not like the British countryside they knew. Native gardens existed alongside exotic plants in Tasmania in the 1840s,[14] and there is evidence of attachment to the Australian landscape by non-Aborigines even before that. Alan Atkinson, the historian of Camden, near Sydney, discovered what he thought was 'the first sign of a white man feeling awe or affection for the created landscape':

> And he who Beauty's might despising,
> Still loves to linger near her bower,
> Will find ere long, 'tis worth the prizing;
> And own with throbbing heart her power.[15]

Atkinson found in this 1827 poem a 'strong sense of place unlikely from the pens and mouths of mere invaders'.[16]

Later Australians, of course, found the land already cleared, so their roots were sunk not by creating the landscape but by living or working in it. The Western Australian writer Elizabeth Jolley thought that her love of what her father called 'scenery' derived in part from the bicycle rides he insisted upon when she was a child.[17] Lucy Turner, who grew up on a property near Mittagong, New South Wales, in the 1970s–80s, believed that belonging was a matter not of boundary fences nor clearing, but the physicality of actually being there:

> Making a house and a home very much my place or my family's place, definitely. I really like places that feel like they've got people in them almost more than places that feel very wild.

> So a ranger's house in the middle of the national park you probably wouldn't want to be interested in living in?

> Oh no, I could, if I could make some kind of area round it my place. I think it's important for people to have places that they feel like they've trod, often and all the time. I suppose it's like, you know, dogs urinating or something, you feel like this place is, because you've put so much of yourself into it or walked over it so many times it's going to protect you rather than fight against you. Or it's going to do something for you, like grow you some food. It's that sort of physical knowledge of the place, isn't it. You've gotta actually got to have had your body there.[18]

SPECIAL PLACES

Margaret Johnson had her own special places which she created by having her body there. In the homestead *the old kitchen had a life of its own,* cats curled up in warm corners, nappies hung on the towel rail, the iron kettle simmered on the hob, *a lovely feminine brittle gum* stood outside the window. Each cup of tea and brownie added to the thousands of other teas and brownies consumed in the same warmth, at the same table, on the same seats. *The daily events and the physical characteristics of the country are all sort of intertwined, it's a really very special precious, strong thing.* Had the walls themselves absorbed something of this human activity? *Oh yes, I think so. There's a patina, a richness.* There were other special places about the homestead which Margaret Johnson observed and absorbed. One was the jasmine bush outside the bedroom window where the birds gathered.

Margaret Johnson had her family soon so her earliest and most intimate attachments were associated with her children at sites near the house and garden. In the early years of her marriage, the play areas about the homestead became her most special places, *the dappled light of the trees and the needles under the pepper tree, and that sort of very peppery-tree smell.* These were the sites of playful imagination. Jim Johnson arranged to have a huge flat-topped granite rock placed close to the house in the front garden. Around it Margaret Johnson planted trees to form a grove. *I used to weave all sorts of stories for the children and grandchildren sitting on the rock.* 'There's Mrs Willy Wagtail, she's got to get a bit of water for her babies up in the bottlebrush. I wonder where Mr Willy Wagtail is?' Sticks, stones, feathers lying about, became the items associated, at the precise moment, with the precise location. *So that's a very favourite spot.* A certain gully was associated with foxes: it became Fox Gully. 'Mr Rat' lived in the vegetable garden: there he was between two bales of hay, cleaning his whiskers, watching the children. Goblin Grotto, *the location of an enormous number of family picnics,* was a natural amphitheatre surrounded by box and kurrajong and *lovely blue granite rocks.* Here lived rabbits, foxes, snakes, lizards and goblins: *it was the sort of place they might be living in.* The Goblin Grotto creatures, each in their different houses and cooking areas, were supposed to have extended families who visited each other, played tricks and had adventures. *You get quite fascinated by those large eyes just looking at you taking in every word as though it's the absolute truth.* Near the front gate was *a wonderful little culvert with willows and poplar trees, and underneath the willows there's almost a house, just absolutely gorgeous.* In this place of silence, leaves fell upon the bank and the water was like a carpet. The culvert became the location of stories about real animals who came to swim or drink. Each occasion Margaret Johnson and the children picnicked there, the culvert held a different story about another animal and another purpose, but always the theme was the events which had occurred in this willowy house, in this shadowy room.

There were wilder and more intimate places, unsocialised and deserted, where Margaret Johnson went alone for the day with the dogs. *There are lovely places to go when you're sort of sick of family and house and home you can just take off and go to those wonderful places ... Your attachments are really secret things. I don't talk much to others, but it helps to keep you on the level ... you need an inner peace.* One special place was the site she selected as her burial place, *a lovely spot, up the creek, under a gorgeous gum tree that comes over a big granite rock and the water runs over it very softly.*

This was the Windermere which Margaret Johnson understood and absorbed over thirty-seven years and from which, in 1993, she was to be separated. Reflecting on these deep attachments, she considered that beyond the special wild places which bore their own intimate character, was the further dimension of socialisation: the events of family life had laid other layers of meaning over the homestead and the land—school, sickness, parties, visitors, celebrations, funerals, marriages, picnics. To the private sites and the public collective spaces had been added, in the magical last decade on the property, an unlooked-for relationship with her grandchildren, who became as intimately involved in the same special places of Windermere as had her children. *The actual rich overlay is when your grandchildren come back, and they've been here, and you've been introducing them to all the things that you've loved. Just showing them the special things at Windermere made them stronger to me.* Grandchildren at Windermere provided *another layer to life.*

How have humans formed attachments to their special places? After marking out boundaries, clearing, working and the physical presence of being, the last part of bonding with the land is, for some, the creation of special, intimate or sacred sites. Individual differences are as great as cultural. The philosopher Karl Jung thought himself 'spread out over the landscape and inside things'; there was nothing, he thought, in his residence, that had not 'grown into its own form over the decades, nothing with which he was not linked'. So in tune was Jung with his special places that one night he witnessed hundreds of peasant boys marching past his home; these he surmised from reading old manuscripts, were the souls of the seventeenth-century soldiers of Wotan's Army. The soldiers marched over the land periodically, visible, Jung believed, only to those sufficiently in tune with the vibrations of the special places.[19]

There are many Australian analogies of deep spiritual bonds with the land. The Aboriginal historian Jackie Huggins returned with her mother Rita to Carnarvon Gorge, Queensland, from which Rita had been removed sixty years earlier.

> The way my mother moved round, kissed the earth and said her prayers will have a lasting effect on my soul and memory because she was paying homage to her ancestors who had passed on long ago but whose presence we could still intensely feel ... The land is our birthing place, our cradle; it offers us connection with the creatures, the trees, the mountains, the rivers and all living things. There are no stories of migration in our dreamtime stories. Our creation stories link us intrinsically to the earth.[20]

The poet David Campbell believed himself to be almost a physical part of the landscape of the southern tablelands of New South

Wales where he lived and worked, his own psyche both creator of, and element in, those significant sites. Campbell was a participant, as a critic put it, in a 'surreal dimension of energy, light and dance':

> The hawk, the hill, the loping hare
> The blue tree and the blue air
> O all the coloured world I see
> And walk upon, are made by me.[21]

The novelist Kate Llewellyn described her discovery—or was it creation?—of a sacred site in the Blue Mountains, near Sydney:

> There is no other access to this place and few people go there. Aboriginal drawings in the rocks lie wearing in the sun and wind, a tribal memory of the fish and whales and dolphins the people once caught here. The whole place has a magical and mystical feel about it and I never go there without being affected. It is secret, it is sacred, it's menacing and takes you in a gulp.[22]

Clare Milner, a farmer from Braidwood, New South Wales, believed that the people of the past shared sites with the living. She heard voices, she encountered and spoke with people evidently from the nineteenth century. Once she saw a slab hut suddenly become visible in an apparently empty paddock. When she investigated, she discovered buried in the grass the ruins of the cottage she had seen:

I was out in that paddock with a friend, and we were actually picking up some sheep that had mysteriously died in that paddock. And I looked up to where Annie was standing, and I suddenly saw a house, a slab cottage with smoke. And there were rabbit traps and harnesses hanging up along the verandah, and I said, 'Oh Annie look, there's a house there', and she of course thought I was mad. And then I ran up to where she was standing, and in amongst the grass we found the foundations of a cottage, and the foundations of a chimney. And that was very extraordinary proof to me that I did feel things that were there.[23]

Since at least the 1930s non-Aborigines have wondered about and compared their feelings of belonging with what they knew of Aboriginal attachments to the landscape. The rather pessimistic school of nationalistic poetry known as the Jindyworobaks looked beyond scenic beauty to form for themselves a spiritual connection with the land. Frequently they failed, because, they believed, the brutal dispossession of the land by earlier generations had severed a possible connection. Rex Ingamells wrote in 'Unknown Land':

> We who are called Australians have no country.[24]

Following the more optimistic middle decades, the poetic theme that non-Aboriginal Australians could not form a legitimate attachment to the land returned in the 1970s. Non-Aborigines had appropriated the land materially, not spiritually: how then could they master the spirit of the place?[25] Judith Wright wrote:

> The love of the land we have invaded, and the guilt of the invasion—have become part of me. It is a haunted country.[26]

When Margaret Johnson speculated with her children about who might previously have lived in Goblin Gully, Aborigines played no part. Reflection came later. Windermere Station was established in the country of the Wiradjuri Aboriginal people, but no one in her time remembers the Wiradjuri working or living on the property. Few artefacts have been found in the paddocks. In the 1990s Margaret Johnson tried to imagine how much the Wiradjuri loved the land:

After [my experiences] I really appreciated what the Aborigines must feel for their country, I mean it must be much stronger in them, because they haven't had the overlay of other things that I've got, another sort of education. They've got this very strong feeling for land.

So have you, though:

Yes, yes, but I can see how they must feel—what are they without that land, whereas I can sort of gather myself together and say, Right, I'll do some course, or other sorts of things at my fingertips.

Most farming children of Margaret Johnson's generation grew up heedless of Aboriginal attachments to the land into which they were sinking their own roots. In the 1980s other reflective farmers, like she, were finding that the values by which they had been raised had ignored the prior or continuing attachments of the indigenous people. Jill Ker Conway grew up at a station a few hundred kilometres from Margaret Johnson's Windermere. At university she asked herself:

> Who were the rightful owners of the land I had always thought to belong to us? I began to wonder about the aboriginal ovens I had played with as a child, and the nardoo stones [for seed grinding] we had so heedlessly trodden upon as we entered and left our home. What had happened to the tribes that once used to hunt over our land?[27]

Aboriginal people grieved over the loss of their special places, and their emotions were compounded by the knowledge that the destruction of sacred sites is followed by social and spiritual decay and death. Gurra, an old man of the Arrernte people of Central Australia, told the anthropologist Ted Strehlow:

> The Ilbalintja soak has been defiled by the hands of white men. Two white men came here to sink a well. They put down into the sacred soaks plugs of gelignite, to blast an opening through the hard rocks to the bottom. But the rock was too hard for them. They had to leave without having been able to shatter it; they took ill soon afterwards and died.
>
> And now the soak has almost gone dry. No longer do men pick up the grass and the weeds and sweep the ground clean around it; no longer do they care for the resting place of Karora. Bushes have grown up on the very edge of the soak, and there is no one to uproot them. The bandicoots have vanished from the tall grass in the mulga thicket. Our young men do no longer care for the traditions of their fathers; and their women bear no children. Soon the men of Ilbalintja will be no more; we shall all sleep in our graves as our forefathers do now.
>
> There is little here for strangers to see; there is no mountain cave here, only a storehouse in a mulga tree. But though the soak has been forsaken by almost all our people, a few of us old men still care for it. It still holds me fast; and I shall tend it while I can: while I live, I shall love to gaze on this ancient soil.[28]

Strehlow noted this story in the 1940s. Now another generation, perhaps of non-Aborigines, may treasure Ilbalintja as its own, sinking its roots into the red earth, attached to this locality as no other while knowing nothing of its recent history. Humans build upon each other's identified localities and make them theirs. Had the resumed South Australian farmland in this passage really been 'abandoned'?

> With a sweeping gesture taking in all 550 square metres of a brand-new garden a young home owner on the northern fringe of Adelaide explained that when he occupied his new house eight months before the interview the garden had been nothing but abandoned farmland.[29]

GENDERED LANDSCAPES

The sociologist Janice Monk reasoned that, since gender was a central element in human experience, it would be surprising if the landscape did not reflect the ideologies that support distinct gender roles and the inequalities of power relations.[30] Some feminist geographers take it as axiomatic that women experience landscape differently because their understanding of space is in ratio to the unequal power relationship which they share with men.[31] If the words 'landscape' or 'home' bear such different meanings to individuals, 'home' may have gender-specific meanings too. A survey of a Melbourne suburb found that when men praised the 'security' of their home they meant financial security, but women meant emotional stability.[32] But *different* associations do not necessarily mean *lesser* attachments. A survey of British households found that, although women did most of the domestic work, they did not necessarily feel differently about their homes: men and women responded more or less equally to the values of affection, family, relaxation or comfort, while about a third of all men and women denied any feelings of attachment to the home.[33] In this book we will discover, more than once, men and women holding different notions about the significance of particular places they have lost; these seem to be most nearly related to the time and emotional energy expended and absorbed at the different sites.

The geographer Jeanne Kay saw nineteenth-century North American women's space confined to the house interior, garden, clothes line, cellar, dairy and poultry house.[34] Margaret Johnson understood Windermere Station best as a place to be a whole person rather than to be a working pastoralist. She preferred to care for her children or work in the garden rather than fix fences. Neither did she know the proper names for all the paddocks, but, unlike Jeanne Kay's rural women, she divided her attachments between homestead and bush. Margaret Johnson, like Miles Franklin, equally bonded with homestead and hill. Franklin wrote of her Brindabella property:

> Blue smoke wreathed hill and hollow like a beauteous veil ... Irrigation had draped the place with beauty, and I stood ankle-deep in clover. Oh, how I loved the old irregularly built house, with here and there a patch of its low iron roof peeping out of a mass of greenery, flowers and fruit—the place where I was born—home.[35]

The image of the farming man and woman bonding with the land unequally may not suit many Australian farming men either. Margaret Johnson's husband Jim was born on Windermere and has worked the paddocks all his life. He is happiest out in the paddocks with a vehicle or a horse, and his dogs. *I can't think of a better place to be.* He understands what other farmers mean when they say that working their land is like being in church. Margaret and Jim Johnson held much of their love for Windermere in tandem. So have many other Australian farming families of fact and fiction. Katharine Susannah Prichard understood shared loving in her novel *Coonardoo,* in which she held valid the landscape attachments of men and women, Aborigines and Europeans. Coonardoo the Aborigine is 'as part of the place as the air and the trees were ... She grew here, as they say. It would kill her to send her away.' Mrs Bessie, the white woman, 'loved every phase' of life at Wytaliba Station, 'every line of the trees, every light and colour of red earth and pale blue sky, dune-grey mulga and white-barked creek gum-trees with their long dark pointed leaves'; Hugh, the owner, had 'attached himself with a stubbornness there was no thwarting or denying'.[36]

LEAVING

Margaret and Jim Johnson always believed that sons or daughters who share a property with their parents can never learn to manage it properly. Even parents who shift out of the homestead into a farm cottage will still be too close. Throughout their married life they planned that when one of their children returned home to manage the property, they would quit Windermere altogether. In 1992, their son Will returned with his wife to manage the property. Margaret Johnson reflected:

We didn't think of it at all for years and years and years. Except in the back of your mind is always the thought that that's the natural thing to do. When the son comes along you sort of move off and let him take over. So I suppose there's that feeling. And also we had seen lots of other families in this area do it either well or not so well, and so we were hoping we would do it well and make the move without creating any waves and make it a good thing for everybody. [If it's not done well] I don't think the person moving off gains from it, I think it's not a dignified thing then. We always feel that it's good for younger people to see some dignity about things, so they don't have any recriminations cast back on them at all, that it is a natural thing to do and that you're happy about it. Sometimes the older people don't go at all and then the younger people don't develop. They remain a son and a daughter instead of a man and a woman.

Jim and Margaret Johnson decided to move to the town of Young, 40 km from Windermere. To be of most assistance to their children and grandchildren, they planned to buy a suburban house in the middle of town. Jim Johnson would continue working on the farm under the direction of his son. He undertook to find an estate agent and Margaret was to do the rest. A house was found. The Johnsons began preparations to leave Windermere Station.

I asked Margaret Johnson if at this point she began consciously to farewell the special sites which she had shared with her children and grandchildren.

Yes I did actually. The special places, for instance: if I was out with one of the grandchildren at the rock, in the front garden that I particularly loved, I used to think, 'Ah dear, what will it be like in town, there won't be any sticks or stones or leaves or birds ...' so I suppose ... yes I did think of that quite a bit ...

It was those sort of things that I missed, or do miss, and still miss, but I don't allow myself to think of it very much. It's not quite a finishing off, we're too young for that, but it's almost like finishing a painting.

Worst of all would be the day when the removal vans arrived:

When it comes to the crunch, when it came to the actual crunch, it was very physical. It was the physical feeling that was the surprising thing. You think you've got it all under control mentally, you've got this reason and that reason, and you've done all the steps. That's all very well, but the actual physical feeling of leaving it was quite hard. You felt all those feelings of despair, physically, in your stomach almost. Because your mind is dwelling on that actual part of it, it's not looking forward.

Once you turn round and look forward, well you're all right, but you do allow yourself a few times when you don't look forward, and you just concentrate on what you're going to miss, and you feel a sort of despair then. You suddenly become aware of what you're leaving behind. And you analyse those feelings which you haven't bothered to analyse before. You know that you've liked the place, and that you've been terribly happy here, and it's where you brought your children up, and you've got all those memories, and those things have built up like a tapestry. And the actual final rich overlay is when your grandchildren come back, and you've been introducing them to all the things that you've loved. That, I've found, added another texture, and made me probably more aware than what I might have been. So you do have quite a few tricky moments or tricky days when you feel very upset.

So you do have quite a few tricky moments or tricky days when you feel very upset. While this voluntary uprooting was being planned Margaret Johnson slept badly. *It was quite trying in the year before we left. We hadn't realised the worry. You're always hoping to do it well.* A visitor who had moved from her own property came to visit and remarked 'It's just the smell of the place that I miss.' Would she miss Windermere as much? Margaret Johnson thought afresh of the special places. What would happen at the granite rock, so humanised and socialised, which she had so particularly loved? The larger items of furniture were to remain—the sideboard, the dining room suite, the sofa, the enormous wardrobe. There would be no room in the new house, and they seemed to belong in the old. *They just had a feeling for the house, it didn't seem right for them to be moved.* She left a particular painting to her son *as a present to the house.*

Moving day. It was 14 July 1993.

Yes, there was such a day, yes there was. Yes well it was sort of a dreaded day, but it was almost like an operation, it was better to get into it, you could sort of feel it coming, and when it came I felt, well this is it, you've got to do it now. Luckily it was so terribly busy. I had everything labelled and I put it in, but it was very busy. So I probably actually never said 'Goodbye' as such, deliberately. I just sort of put my head down and went, and it was a fairly awful day, actually.

Though she would return to Windermere frequently, she would not feel the same way about it again. Windermere, from 14 July, was Margaret Johnson's lost country.

Wasta Est—the land is laid waste—is a common phrase in the Domesday Book which listed the population and living areas of England after the Norman conquest. Encroachment by the sea, the plague and climatic change put an end to hundreds more English villages by 1400. The greatest depopulator of English farms and villages in England, though, was the phenomenon known as enclosure, by which hundreds of thousands of hectares of arable land were turned to pasture by their owners. In a process which lasted over 300 years, thousands of fields were abandoned, towns emptied, churches deserted and inhabitants uprooted.

One of the earliest recorded enclosures was in 1489, when Thomas Twyford evicted the inhabitants of seven cottages. The jurors called to investigate declared that Twyford had:

> wilfully allowed the houses to fall into ruin and turned the fields from cultivation to be a feeding place for brute animals. Eighty people who worked here went sorrowfully away to idleness; to drag out a miserable life and—truthfully—so to die in misery.[37]

A historian of English lost villages, Maurice Beresford, cited other reports of people evicted from their homes and villages, who 'lead unhappy lives and truly have died in a pitiful state', or who 'left their houses weeping and became unemployed and finally, as we suppose, died in poverty and so ended their days'.[38]

Forcible dispossession was not confined to England or the sixteenth century. Within the last half-century, whole villages in Britain have been resumed or abandoned through military rezoning, new airports, the decline of regional industries or reservoir construction. The twentieth-century poet Gillian Clarke watched a Welsh village disappear under the waters of a reservoir and wrote:

The people came out in pairs,
Old, most of them, holding their places
Close to the very last minute,
Even planting the beans as usual
That year, grown at last accustomed
To the pulse of the bulldozers
High in those uphill gardens, scarlet
Beanflowers blazed after
The water rose in the throats of the farm.

Only the rooted things stayed:
The wasted hay, the drowned
Dog roses, the farms, their kitchens silted
With their own stoves, hedges
And walls a thousand years old.
And the mountains, in a head-collar
Of flood, observe a desolation
They'd grown used to before the coming
Of the wall-makers. Language
Crumbles to wind and bird-call.[39]

The Appalachian poet Jim Wayne Miller ironically described the disappearance of farms in Tennessee:

Sometimes a whole farm family comes awake
in a close dark place over a motor's hum
to find their farm's been rolled up like a rug
with them inside it. They will be shaken onto the streets of
Cincinnati, Daytona or Detroit.

...

Divers searching for a stolen car
on the floor of an Army Corps of Engineers'
impoundment, discovered a roadbed, a silo, a watering trough
and the foundations of a dairy barn.
Efforts to raise the farm proved unsuccessful.
A number of Tennessee farms were traced
to a land-developer's safe-deposit box

in mid-state bank after a bank official
entered the vault to investigate roosters
crowing and cows bawling inside the box.

...[40]

There are, perhaps, proportionally as many Australian lost communities as there are English or American. For every lost town like Leigh Creek in South Australia (demolished to enlarge a coal mine) there are a hundred lost and forgotten communities of tumbledown houses, overgrown camp sites, broken bricks, disused axe-grinding grooves, foundations of Aboriginal mission stations, apricot and plum trees in the midst of paddocks, wild irises at abandoned railway sidings, gold mines, crossroads, river crossings, and sailing-ship ports. Some of the inhabitants probably left each of these loved places actually or metaphorically 'weeping and became unemployed'.

Lost and dead places can be as small as a suburban garden. A Melbourne pensioner who had moved to an old people's hostel described what he had been forced to abandon:

> But I had my own garden. I had a beautiful garden, all beautiful plants, which I had to give away ... You find—and I guarantee that if you interviewed other people, they would say the same—they miss their gardens. There was people who, when they came in here, all they did was walk around crying all day because it was part of their life. Well with me, I felt as if my heart was torn out.[41]

Lost places can be loved by many people. An entire community was destroyed in the Burragorang Valley, near Sydney, in 1957. Hectares of farmland were flooded and 170 families were evicted by resumption for the Warragamba Dam. Desperate efforts to save the valley were made by community organisations and individuals.[42] A long-time resident of Burragorang Valley, Cecil Pearce, wrote:

> To us children it was like a very vivid nightmare. For us to have our farm and home covered with water, to have to leave The Valley and live elsewhere, was something unbearable even to contemplate ...
>
> We slowly but surely began to feel the effects as their [the Water Board's] inspectors began to enforce laws which prevented us from

> running our pigs on the river flats. The Wollondilly Shire was reluctant to spend money on keeping the roads and bridges in repair. The farmers began to adopt a similar attitude. Why erect new fences and sheds or spend money on other improvements when the farm was to be resumed in the near future? ... The result was an overall deterioration in all that mattered and slowly and surely The Valley took on an appearance of neglect and desolation.[43]

Protest was vain. By 1957, as a local poet put it:

> The water's over Nattai Bridge
> The last mail has been run
> And lonely Kiaramba Ridge
> Glows in the setting sun.[44]

Owen Pearce, Cecil's grandson, commented on the people's exit from the Burragorang Valley:

> My grandfather talked a lot. 'Don't worry about it girlie, I'm not going peacefully. If they flood The Valley, I'll float out on a log' ... Of course, things didn't happen in that manner. The Valley people just went quietly and sadly ... gave up their farms and moved on, so that Sydney could quench its thirst.[45]

Left their houses weeping and became unemployed.

RETURNING

Jim Johnson continued to work at Windermere while sleeping in town. *I'm not worried about whether it's right or not: it's been done.* He noticed the greatest effect of their absence from the property in the dogs. *That worries me more than the human side, the humans can solve their own problems, the dogs can't.*

Margaret Johnson reflected on the first few weeks in Young:

I don't think the first week was too bad, I think it was a few weeks after that were pretty awful. I found that really people said 'Oh dear, you've moved quite well' but I think you do keep a lot of things to yourself, and you're fairly bright and cheerful because after all, we were moving to a very nice house, and our family was still here.

But in actual fact I had to get up very early in the morning, because if I woke up and stayed in bed and thought—that was the end. That was a very dreary beginning to the day. Physically it was very difficult, rather than mentally, because your mental processes tell you that it's perfectly all right and not to be ridiculous, and it's a nice sunny day, and to get going, there's a lot going for you. But actually physically, I think it's, I think it's just that tie to the spirit of the place. That was very hard.

Did the family know of her feelings? *Maybe they did and maybe they didn't.* In town, though she was surrounded by children and grandchildren, life seemed to have less purpose. The loss of Windermere seemed like a recently closed wound—tender, painful, easy to re-open.

Returning to Windermere brought pleasure and pain. Pleasure: when she was able to return, not as a mother or grandmother, *but just for my own sake*; it was *just wonderful*. There again was the granite rock, *the dappled sunlight falling on the porch, the lovely feeling of relaxation. I just feel quite at home and I think the country notices [me]. I say hello to the creek and hello generally speaking, and hello to a few clouds. They are the things I just adore out there, because there are clouds* [in Young] *but they are not nearly as intimate as the clouds out there, whether they come down between the hills and you just seem closer to them; but they're really quite spectacular.* She visited the old quiet spots, *not sad, but very much at home*. On one occasion she avoided the homestead and kept to the creek with her grandchildren. *Lovely cold sand just under the running water. We were soaking it in … as if we'd never left.*

Pain: *It's still mine but it belongs to someone else.* The first time Margaret Johnson entered the homestead after moving to Young she went upstairs and burst into tears. *Crying for that loved one, that old house.* Would it have been easier to leave and never to have returned? *At least you've got the memory intact, you don't have to front up to it again.* Memory was like the death of a loved person. *Just when your feelings were under control this can take you by surprise, there's a real physical tie to the land and a feeling that is part of your spirit that's divorced from all arguments of logic and reason and behaviour, just something that assails you.* She behaved, she thought, quite well on her visits, but when she left she felt *down in the dumps*. The exercise had taught her about life, she reflected. *I can accept death better now. A little bit of you dies, not for the material things, but for the spirit of the place.*

We are far from the professional meditations of the rhyming ruinists who in eighteenth-century Britain delighted in evoking nostalgia and the littleness of human ambitions which they imagined crumbling like the ruined castles of their contemplation:

> This princely pile, with all its splendid spoils
> Sinks midst the havoc of intestine broils
> In prostrate ruins lost, and dark oblivion laid.[46]

Margaret Johnson's experiences mirrored those of many others returning to their lost places. The greater the emotional investment in a site, generally the more painful is the return. The psychologist Marc Fried, studying the forced removal of people from the west end suburbs of Boston, was confronted by their grief, sense of painful loss, continued longing, depression, helplessness, direct and displaced anger and idealisations of the lost place. Grieving for the west end was 'a widespread and serious social phenomenon following in the wake of urban dislocation'. Why such passion, why such despair? Because, Fried reasoned, an individual's sense of continuity of person and community was intimately bound with spatial identity.[47] It does not take a psychologist to sense the pain of loss and dislocation of resumed city blocks, rural properties or suburban homes.

Returning to special places can be as traumatic as living apart from them. Jim Hall, a miner, returned to Marwood in eastern Tasmania and recalled:

> Well we were there for nearly forty years, and when we left she [Mrs Hall] broke down and cried.
>
> And now you'd hardly know there'd been a mine at Storey. Where the old compressor used to be there's blackberries as high as this house. I went out there and had a look, but I'll never go back again.[48]

Rita Huggins returned with her daughter Jackie to Barcudgel Station, Queensland, where she had worked many years before as a domestic:

> When I got to the place, I was saddened by what I saw. The station house had been burned down by bushfire some years earlier. There were a few things still among the ruins. I found an old mincer that I

> probably had held in my working hands. The old frame of the shearing sheds still stood. I looked around me at these ruins and tears welled up in my eyes ... I saw my life pass before me on that day, at that special place.[49]

A Burragorang resident, Laurence Brunero, returned to the Warragamba Dam lookout:

> Some scenes from my childhood memories are now hidden from view ... a paradise lost beneath the waters of the Warragamba Dam. Many times I have returned to the lookout above Oakdale and gazed at what is left of the Burragorang valley. As folk around me wonder at the magnificence of the mountains and the lake, my thoughts are of times and events buried forever beneath the waters.[50]

So many lost places; so much pain. Yet some Australian writers, evidently confusing local attachments with a more problematic national identity, continue to assert that Australians have no sense of belonging to place. The literary critic Robert Dessaix has written 'The hunt's exciting and there are stories to be told about it, but no hearth to tell them around, over and over again.'[51]

It is not only Australians who have not understood, or turned their eyes from, profound attachments to the sites of lost homes, towns and countries. Christianity, a proselytising and formerly peripatetic religion, transports its sacred objects: a church can be erected anywhere, and the site may be deconsecrated as well as it can be sacralised. Capitalism holds that land, like labour, is a commodity to be bought and sold; emotional attachment is irrelevant.

A third reason why we have underestimated the power of a place to affect those linked to it shifts the focus from humans to the sites themselves. Social scientists have pondered the declining importance of studies of particular towns or areas, which in the 1950s and 1960s were the standard literature of geography and sociology journals. They reasoned that in the 1970s social scientists, attempting to understand the collective workings of human society, focused on the more nebulous concepts of 'community' or 'class'; as traditional rural communities began to decline, their sense of belonging was set aside as unimportant, as if the collective feelings of people about a town were the same as individual feelings about individual sites within it. Anthropologists working in the third world understood the importance of 'place' among indigenous or colonised peoples; even nineteenth-century European peoples were understood to have been rooted in exact locations. These attachments became associated with parochialism and localism, and became early casualties on the perceived continuum of progress, out of temper with the nation–society focus of most contemporary social science.[52] But place, in the end, is where human events take place. The post-modernist geographer Edward Soja has distinguished 'place' from 'time' in the explanation of why the world is as it is. Soja held that for two centuries historians and sociologists followed too closely historical time, the 'what happened next?' syndrome, as both explanation and context. Historians treated place as 'fixed, dead, undialectical', while time was held to be 'richness, life, dialectic, the revealing context for critical theorization'.[53] It was appropriate, Soja believed, for social scientists to return to the lost matrix of place as context and *sine qua non* of human activities.

A year after leaving Windermere Station Margaret Johnson felt more comfortable in Young. She and her husband had laboured to make themselves belong in a new site, and Margaret's efforts had been rewarded by commitments like evening classes and a schedule of minding her grandchildren. If she went to Windermere for a day visit, she might ring to say she was coming, but did not feel the need to have to stop for a cup of tea. She and Jim Johnson also stayed a night or two at Windermere in a vacant cottage. *It feeds the soul again.* She was content with all their actions; had they decided to remain living permanently at Windermere in the cottage, they would only have half-left. Now the young ones had established themselves, and she admired what they had done. They had remodelled the homestead: it was necessary for them *to put a stamp on it*. It was good for the children to see their parents acting freely and with dignity. *Only there are moments which you snatch for yourself.* Margaret Johnson did not feel that Windermere was as much hers as it once had been. *You were lent this land and this experience and you should be prepared to move on. You have to let it go.* She walked in the paddocks. Here she could *commune without words, notice where I am, heave a sigh 'Oh this is lovely'*.

This book is about some of the lost places of Australia and the attachments of the people who loved them. All the painful themes which we have touched in this chapter will be revisited. The son of Polish Jews will return to trace, and find, a personal significance in the sites of the Holocaust. An Australian Croatian will return to his family home of 600 years to find it destroyed during the civil war. A high country sheep man will compose poetry in remembrance of his lost beloved mountain home. Houses will be sold, or lost by bushfire or accident. People will return to discover demolishers tearing their loved homes to pieces. Aboriginal adults will return to sites of their childhood they had thought lost forever. Whole towns will be inundated by a reservoir or dug away by mechanical dredgers. Lake Pedder will be drowned and mourned, and Darwin will be blown away. There will be plays, pageants, historical tours, government investigations, signs, exhibitions, books, poems and speeches over lost and loved sites; there will be silences, and absence of signs, and names will be changed on maps deliberately to obliterate memories. Farmhouses will be built on Aboriginal sacred sites, carparks will be asphalted over farmhouses, high-rise office towers will demolish whole streets, and freeways will drive through sacred sites, farmhouses, homes, carparks, streets, factories, cemeteries and high-rise office towers. Dozens of people will seek to articulate what burns fiercely inside them:

> For no clear reason we ache
> To shape of imprecision certain knowledge;
> But things won't integrate, though birds glide
> Like thoughts at the fringes of vision's unshaped words.[54]

While some thoughts will, and should, be felt and not articulated, continuing to glide about at the edge of understanding, this book attempts to shape that imprecise but certain knowledge.

CHAPTER 2 VANISHED HOMELANDS

In August 1993 Luka Prkan left Canberra to return to his birth country of Croatia.[1] He had emigrated in the 1960s, and had returned to Yugoslavia with pleasure several times, but the civil war of the 1990s made this homecoming agonising. He made his way to his birthplace, the family home for 600 years, to find it in ruins, rubble, block on shattered block. Pieces of window, half a door, a twisted iron bedframe. Half the outside staircase pointed to nothing. The enormous stove which had kept his family warm when he was a child had vanished. Donkeys, turkeys, geese, pigs, all the farm animals slaughtered; other once-domesticated animals charged past in terror. The ruin seemed so small and pitiful. Lost homelands are dead localities.

Lost countries are dead people. Luka Prkan's father died in the 1970s and his brother was wounded by a mortar splinter during the civil war. His mother, who had somehow survived the mortar attack on her home, had at first refused to leave the ruin. He found her in the refugee camp to which she had been escorted. He was not sure whether she knew him or not. It was said that from time to time she scrambled back to the wreckage of the family home, to be extricated from the stone fragments and returned to the refugee camp.

A lost country is a dead culture. Luka Prkan's father had supplied wooden wine barrels to the district. He cut the oak in the forests, shaped it, bent it over a cherrywood fire, forged the iron bands and exchanged the barrels for produce. Luka Prkan thought of his father bending over his work, the workshop warm and secure, prosciutto above, a glass of grappa nearby. The older style of living was all but gone by the time his father died, and the civil war put an end to every trace of it. Where was the old Croatia of the *doma*, the households where he was always welcome for a meal or a glass of home-pressed wine, the farm animals sharing the winter living spaces, the simplicity of material life, the beautiful coastal scenery? Where was *Korzo Riva*, the evening promenade when people walk the pavements, arm in arm, a coffee here, a slivovitz there, a meeting with friends in the piazza and in the bars, a leisurely conversation, a game of cards or a heated political discussion, the feast day—the intimacy of life in which every place, every person was important?[2]

Lost countries are the terrible events which occur in them. The church of St Ilija where his father was buried was too dangerous to visit: a relation told him that the churchyard was almost certainly destroyed during the bombardment. As Luka Prkan travelled through his almost unrecognisable homeland, no one but he seemed to take much notice of the bombs and mortar shells. Far from the ecstatic welcome expected by a long-absent traveller, Luka Prkan felt anger directed at him, especially by homeless refugees, that he was from the west and ought to be doing more to help. He felt angry, useless, ashamed. He was taken for a stranger, but why? He wanted to help, but how?

I think generations are going to be destroyed and distorted from now on. Lots of kids, you can imagine, without parents. This is the worst war in man's memories. You have no idea. They're burning kids alive, roasting them, picking eyes out, it's so completely sadistic, incredible. The stories I heard I just can't believe it. One of my nephews was on the front line. When you look at him it's not him any more. He's absent-minded, he talks to himself, shaking. He said with a raised voice, 'You know nothing, Nothing!' ... I tried to have an opinion—'You know nothing about it'. Quite angrily. And rightly so.

Luka Prkan returned, shaken and demoralised, to Australia. In peaceful Canberra the birds sang, friends came for lunch, the plum trees dropped fruit, the dog barked, olives and wine lay on the table. The violence and despair of his birth country seemed far away, and even further was the birthplace that Luka Prkan once knew. Croatia was a lost land.

MIGRANTS OR EXILES

Some migrants who come voluntarily to Australia seek to re-establish the old land in the new. The archetypal Dutch male migrant of the 1950s was a dairy farmer or market gardener while the archetypal female re-created Holland inside the house. A typical Dutch-Australian house, according to the sociologist Jerzy Zubrzycki, was neat, trim and tidy, with dark-coloured oak furniture, a tapestry over the fireplace, teaspoons in a cup on the sideboard, Delft-blue china, a cuckoo clock, small brass or copper objects placed round the room, and souvenirs with the name of the home town in Holland engraved on them.[3] Italians also formed enduring groups, first of 'Little Italies', then of larger communities. At Little River, Victoria, pre-Second World War Italians used to visit their neighbours on winter evenings. They gathered in warm farmhouse kitchens at *filo*. In the recollection of Rino Baggio, the men talked about farming and the women 'what women talk about'. At about 10 p.m., *vino brule* was served. Baggio wrote: 'by dint of listening to the grown-ups talking about the places and the people they had left behind, I felt like I had been there myself, and knew every twist and foible of their missing friends' personalities.' During World War II the Little River Italians held dances—of all places, in the Returned Servicemen's Hall. For half the night the community would dance traditional steps, and a 'tall moustachioed Tuscan man and his chignoned wife', both former professional entertainers, would lead off with 'Vivere' and 'Mamma'.[4] The community culture and attachments of the home country—or home town—is what Luka Prkan could find neither in Australia nor Croatia in 1993; it was what Margaret Johnson looked for in Young.

For many years an army of sociologists studied how, why and if first generation migrants ever really belonged in new countries. They studied personal and social background, motivation, expectations and the nature of the receiving society[5] and developed a number of theoretical models about how assimilation occurs. A hypothesis of the 1950s described the first generation of immigrants clinging to their first culture and loyalties and trying to inculcate these in their children.[6] Such mechanistic models, which detected stages like 'naturalisation', 'absorption', 'assimilation' and 'acculturation' were replaced by more sophisticated theories which allowed for individual difference, changing attitudes throughout the whole of life, and the stability of a national or regional culture from decade to decade and generation to generation.[7] Immigration theorists now allow that the process of belonging in a new land is much more complex than previously imagined.

A special group of migrants cannot return because their homeland, like Yugoslavia, no longer exists, or because, as in Vietnam, they might be imprisoned. Two hundred thousand displaced persons came from Europe to Australia between 1947 and 1953, and the establishment of communist governments in states like East Germany and Czechoslovakia made return impossible. Refugees from these nations carried the burdens of the true exile. They understood that the land, the special sites, the people and the culture of the old country were forever inaccessible; they were transported to, and tried to set their roots in, an alien culture which also seemed ultimately inaccessible. Yet the lost homelands are ambiguous. To sons and daughters, lost landscapes bear upon them and within them the vibrations of the atrocities and the sunnier days before the darkness. The writer Andrew Riemer, recalling the unhappy experience of his Hungarian refugee parents in Sydney, reflected that the 'tragedy of such lives is ... an inevitable and natural nostalgia, an ever-present clog at the exile's heels'.[8]

One of the few Australian sociologists to make a special study of exiles was Jean Martin, who in 1955 worked with a small group of recently arrived displaced persons. She described them as 'an unusually insecure and suspicious group'. Most, she wrote, 'had lost close relatives during the [Second World] war; they had left their own home unwillingly, suffered humiliation and privation, and after the end of hostilities, experienced years of empty, idle uncertainty before their future was finally decided by resettlement overseas.'[9] Martin noted the complexities of psychological adjustment overlooked in the mass-distribution standard questionnaires, that refugees whom immigration authorities thought well assimilated, for example, in fact were demonstrating conformist behaviour 'symptomatic of a disturbed and insecure personality'.[10] Cultural differences were also overlooked. Vietnamese Chinese, familiar with business and trade, accustomed to being in a minority and having an effective clan and family network, were more likely to accept and be accepted in Australia than ethnic Vietnamese themselves.[11]

In the new country dead homelands remain moving shadows throughout the whole of life. Jean Martin understood that. She realised that the behaviour of the displaced persons whom she studied was based in part on:

> resistance against the violation of the individual's personal and cultural integrity, the attempt (not necessarily successful) to re-establish a stable existence, ordered in terms intelligible to past experience, and—in a sense encompassing these first two themes—partly deliberate and partly unacknowledged gravitation towards a milieu where the minimum degree of fundamental personality change would be required ... the choices they have made reveal the working out of needs and aspirations well established long before these people knew that they would be called upon to remake their lives in a new homeland.[12]

Many Jewish survivors of Nazi Europe have not returned to their homelands and have no wish to. The sociologist Naomi Rosh White interviewed eleven Polish Holocaust survivors now living in Australia, of whom ten wished to be 'as far away as possible from the war and its memories'. Most of the ghetto or concentration camp survivors were emotionally empty. 'Rancia' was never so unhappy as in her first years in Australia. *The war had burnt me out. I was so burnt out that I felt I could not give to my child.* Others believed that Europe itself was finished. *The Nazis taught me that the Jews have no country.* But those genuinely determined to belong in a new country forced themselves to do so. *One is left with a choice. Do you want to die of your pain, or go insane because of it. Or do you want to resolve it somehow?*[13] 'Hania', a Polish concentration camp survivor, forced herself to remember:

> ... the picture in my mind of Auschwitz and Treblinka is a picture I look into everyday. I see it. And it keeps the pain alive. That's what I want. I don't forget and I can't forget what happened. It's a part of my life, something I went through. I don't want to forget. I want to imagine how my sisters felt at that moment when they had to give up their children or to go with them in their arms.[14]

It is not uncommon for such people who have rejected their birth countries to describe themselves as European in origin rather than Croatian, Polish or Russian.

One exile who set herself to belong afresh in a new homeland was Susan Karas. She was born in 1920 and grew up in the town of

Zilina in the Tatra Mountains of Slovakia. Members of her family, with other citizens of the Austro-Hungarian empire before the Second World War, felt a complex sense of patriotism, a dual citizenship—local and Viennese—which transcended national and racial diversity. The new nation of Czechoslovakia enjoyed a brief period of national optimism after 1918, and inculcated patriotism and a shared identity successfully enough for some of its citizens to accept, in another hemisphere and another age, the notion of equality before the law rather than through ethnic affiliation. Czech, Dutch and even German Jews were likely to have described themselves before 1939 as citizens of their birth country first and Jewish second.[15]

Susan Karas saw her parents and sister for the last time in 1942, when they were transported to Auschwitz in a cattle truck. She herself was sent there in 1944, but somehow survived and migrated from Czechoslovakia to Australia with her husband in 1949. *There was a mass psychosis of leaving. Everyone was.* She has not returned to what she still regards as tainted country. *What kind of feeling will I have if I go back there? ... They look into your face and look into your eyes like it's nothing.* She has visited Europe twice, but has never come closer than half an hour's bus ride to the Czech border.

How could I ever arrive at the railway station again? ... What kind of feeling do you have when you are vis à vis murderers? ... Deep down I would love to go back for a little peek, but something is stopping me. The whole of Slovakia is stopping me ... Czechoslovakia means no more to me than Switzerland ... I have lovely childhood memories but it's finished: it's done with.

It was no accident that by 1960 Australia was Susan Karas's first love, nor that she mentally kissed the soil of her adopted country whenever she returned to it. In about 1955, after five years of acute homesickness, she made a conscious effort to start her life again as a citizen of a country she thought worthy of attachment. Susan Karas abandoned her Jewish religion as meaningless, though she later returned to it as a non-observer. In the Sydney suburb of Double Bay she tried to avoid other Central Europeans. Laboriously she made herself learn to play bridge and to drink milk in her tea. She insisted that her daughter Sylvia learn English, served her baked beans, devon sausage and cheddar cheese, and sent her to Presbyterian scripture classes. She was ready to answer questions about her past though Sylvia respected the silences. There was no need to ask for details about her mother's forty-eight hours without food or water in the cattle truck on the way to Auschwitz.

Many refugees find that the emotional soil of new countries is shallower than their homeland's. Susan Karas willed herself to sink Australian roots, but the lost country was not easily obliterated. Every year in the 1950s and 1960s the Karas family enjoyed the Blue Mountains, near Sydney, for its cooler climate, its pines and its firs. *We love it, in the mountains, surrounded by those mountains which we loved, where we grew up.* Sylvia sensed her mother yearning for something which could not exist again because she had willed it so. Yet to this day, each Saturday, Susan Karas meets two women from the town of Zilina where she was born more than seventy years ago. They catch a ferry to Circular Quay, have a coffee and *sacher torte*, walk about talking of *everything and nothing, of old times as well as other things. Do you remember, do you remember, do you remember—every Saturday.* Yes, Susan Karas would love to revisit the fragmentary sites of her childhood, where on summer Sundays, so long ago, she would picnic and gather mushrooms on the slopes of the Tatra Mountains. Nearly half a century ago she engraved the names of forty-eight members of her immediate family on a plaque on her grandmother's tomb. Perhaps, overgrown in the Zilina cemetery, she might find it again. *The rest of the family have no resting place because they were gassed in the gas chambers. No one has seen it ever since. I'd love to see it once more.* But the land in which almost every member of her family perished has ceased to exist. *I have lovely childhood memories: now it's finished. It's done with.* At her death her body must remain in Australia. *Nothing of me should go back!* Susan Karas has learned not to remember, she has willed herself to forget. *If I met you tomorrow I would not remember your face.* Yet every night, as she brushes her teeth, she mentally recites the names of her forty-eight nearest kin who lost their lives in the dead land of terrible memory.[16]

LEAVING A LOST HOMELAND

Irena Petrovna Sinelnikova was born in Russia in 1919. Her daughter and son-in-law exiled themselves from their native land for political and cultural reasons by simply not returning from abroad. Irena

Petrovna's reasons for leaving were less complex. At seventy-three years of age she fled her native land to join her daughter, who was her sole blood relative. She simply shut the front door of her St Petersburg home in 1992 and left, telling her neighbours that she would soon return. At the airport she said to Visa Control that she was going on a holiday: the most precious of a lifetime's possessions she crammed into a single carrier bag. *If I return there I'll be absolutely alone. Friends are not your people, not your family.*

The new land of Australia was strange, and the Canberra suburb amid rolling paddocks where her daughter lived in 1993 was stranger still. Until Irena Petrovna's exile, she had lived and worked in St Petersburg and does not remember, in all her life, ever eating a meal alone. She survived the siege of Leningrad and still recalls the wondrous patterns of ice in that winter of bitter starvation. Now her homeland, her home, her possessions, her friends, all her past are irretrievably lost. In her St Petersburg home, behind the drawn curtains, the grand piano and an antique violin, the precious library, photographs and mementos, stranded and meaningless without their human associations, gather dust and silence. Neighbours, unaware that Irena Petrovna will never return, still come to water her garden. Irena Petrovna believes that the Russian state will seize her possessions. Is it painful for her to think of these things? *Yes, at first, but the longer you stay here, the less you need them.* Redecorating her daughter's lounge, she reaches for the tape-measure, and remembers it is in St Petersburg. She misses her Russian culture, and the forests. As if still at home, she imagines the cathedral and the bridges spread out before the French windows. *It was so beautiful.* She misses Russia much more than does her daughter. *Of course I would like to see my friends ... to feel at my own home, but I see that it's all illusions, illusions. I cannot live alone.* While her daughter has nightmares of raids by the secret police or last-minute border apprehension, Irena Petrovna dreams of the gracious avenues of her city, her friends, holidays on the shores of the Black Sea, and the forests of Estonia, *places you will never forget ... Even when I close my eyes, awake, I can see it.* She wants to be buried in Australia, *in a very beautiful place,* where there will be someone to place flowers on the grave, so that she will not be alone. In public places Irena Petrovna speaks in English. *We don't associate with Russians here. None of the family sends photos back to Russia, and their news is brief. Why should we give the KGB a free photo file on us?* In her daughter's home she talks to her grandchildren in Russian, they respond in English.[17]

Seventy-five thousand South-East Asians, especially Vietnamese, shared a similar experience to Irena Petrovna. Mostly less than twenty-five years old they arrived in the first ten years after the fall of Saigon in 1975. Many had endured horrors in the boat passage or refugee camps worse than those which had caused them to leave Vietnam. Among their reasons for escape were fear of government, blacklisted, property confiscated and re-education camp. Their traumas were little different from those experienced in Nazi Europe: 102 heads of households evacuated from a South Vietnamese battlefield displayed a higher level of emotional disturbance than any group to whom they could be compared except prisoners of war.[18] Yet few Vietnamese rejected their homeland as comprehensively as many Jewish Europeans. Almost all the dozens of Vietnamese interviewed in Footscray, Melbourne, in 1984 were deeply homesick and wished to return when it was politically possible.[19] 'Nguyen N.' told the sociologist Lesleyanne Hawthorne:

> Now I live as if I am living in a dream. I feel as if I have freedom, and every material thing I could wish for. I was lucky to be allowed to choose Australia as my second home. But I feel I love my people, my family, everyone who stayed behind ... Every time I see an Australian child holding a bottle of milk and drinking, I feel misery for the children of my country who are living without any good things ... My life now is that of a machine ... There is nothing interesting in my life now. Yet I feel I lead the life of a little king, the kind of life full of plenty I never thought to have again ... To me, life in Australia is just the four walls of a house.[20]

We have our freedom, but no happiness was a common lament in the first few years of such loneliness, incomprehension and rejection.

One common turning-point in the life of the exile is the death of parents in the home country. A second is the birth of children. 'When my child was born, she was my mother, my father, my brother, all the hopes, all that could have been and wasn't.'[21] Many women seem to become more secure than their menfolk in the country which becomes the home and native land of their children and grandchildren. Before the 1980s the reverse seemed to be true because women were—or were portrayed to be—more unhappy than male migrants. The literature contained many examples. Judah Waten's father, a Russian Jew, wanted to stay in Australia, but his mother did not. 'Even if you make money we must leave this country. We mustn't lose ourselves here. We should only be the living dead in a graveyard.' Waten's mother, in her son's perception, stayed close to her few family members, spoke little English, and talked to Russians about Russia.[22] A later example is Mena Abdullah's Punjabi family, who farmed on the Gwydir in New South Wales in the 1950s. She reported the view held by Punjabi-Australian men about their wives:

> We have our farms. We understand this country. We work with the earth. We know it and it's ours. We have white men as friends. But our women have no such owning, no such friends. This is not their country, and its ways are not their ways. They live only in their houses, only with their children.[23]

Jill Matthews, in *Good and Mad Women*, followed the implications: the migrant woman, expected to provide the security of a stable home while her children grew up and her breadwinner became assimilated, would in the end be left lonely, friendless and unassimilated. The problem was known to the authorities, Matthews commented, but considered to be insoluble.[24] Reality was simpler. Many women migrants stayed home not to avoid the outside world but to look after their children; when the children left home so did they. In the workplace they learned the confidence and the language skills they had previously been denied. It seems likely that the model of the unassimilated female insider contrasted to the assimilated male outsider was true only for the first decade or so after arrival. Susan Karas and Irena Petrovna, if interviewed a year after their arrival, doubtless would both have appeared despairing and alienated.

Cam Trang Dinh was a student during the Vietnam War.[25] A Buddhist, she cheered the progress of the communists fighting their way to Saigon and took part in demonstrations against the Americans and Australians. She supported communism but found, after the fall of Saigon in April 1975, that *reality is not presented*. Her life deteriorated rapidly. Her husband, an officer in the South Vietnamese Army, went to a re-education camp for three years. As the wife of a former enemy, Cam Trang Dinh was prohibited from practising her religion and from further study, and her family booklet, which would have allowed her access to food, education and a better job, was withheld. *Some time the communist police came to our house without permission, go through from front door to back door, look around and what we were doing, what are we eating—even what you were eating!* She became a teacher, but resented having to mouth communist propaganda to schoolchildren. *I feel ashamed when I had to teach something that does not come from my mind.* To make extra money she grew and sold cabbages. In a tiny Saigon flat she lived with, and supported, her mother and mother-in-law as well as her children. She decided to escape because she thought that the needs of her children should be considered above her own: her children had no future in their country of birth.

During 1978–80 Cam Trang Dinh made two unsuccessful attempts to escape, which resulted only in the theft of her passage money. *We were in prison many times and we lost all of our property.* Her two elder sons escaped first: one was lost overboard on the dangerous boat crossing to Indonesia. Her next attempt to escape resulted in her and her youngest son's capture and three months in a horrifying jail. In 1987 she moved to the coast and disguised her city-bred appearance by dressing herself and her son as fisherfolk for six months. Now unremarked by the river police, she joined a secret collective to buy an ancient boat, and after a dangerous passage, in which they bribed the crew of a Vietnamese patrol boat with rings and jewellery, she escaped to an Indonesian refugee camp. Cam Trang Dinh arrived in Australia in 1989 with one son, Tranh Nguyen, and the clothes she stood in.

In 1994 she lived in Eastlakes, Sydney. Though she mourned her native land, she had no trust in the Vietnamese government, no desire to visit the country under the current amnesty. She missed, she said, her friends, family, lifestyle and culture:

I remember the place I live before and the relatives still left behind and sometimes I have nightmares with communist police because they separate us. My son, he was in prison. We lived together. They separate him from me. We are five person together and they separate us, and I don't have permission to look after them. So after two and a half months, when we were released, they all sick because lack of food and hygiene. So it was very awful for them. Sometime I dream I was caught when I escaped. Sometime I dream about my son who was missing on the sea. He's back in memory.

While her two sons remain in Australia so must she, despite its alien cultural practices. *If they feel they are happy here, that their place is here, I'd like them to remember that they are Vietnamese.* Should her children marry Vietnamese? *Yes!* When she has grandchildren she will talk to them, in Vietnamese, of her homeland and its culture. *Their future is very important to us.* Spiritually and emotionally the exile Cam Trang Dinh belongs in Vietnam.

There's something that lives inside us and we cannot break it. Maybe in a long time it lessens, but now it's still there. My past and my family past. It's hard to forget everything. Grandfather and grandmother feel that they are lonely, because in our country the tie between grandfather grandmother and grandchildren is very very important. But now everything is break up. So many Vietnamese people feel unhappy here.

Thanh, the youngest son, agrees: *My country is Vietnam. I want to be in my country but I can't live under the communist control. I remember, every night I saw the police squad around my home.*

Cam Trang Dinh, yearning for her homeland, will stay in Australia as long as her children stay. What should happen to her remains when she dies? They should stay *wherever my children live.*

Male refugees, having spent their working lives apparently putting down their Australian roots, may become less secure as they age. Mourning or reliving a lost past was recognised by Jean Martin as an individual experience which may narrow a migrant's interests or 'push his mind back to the past'.[26] Less imaginative researchers have studied 'regression' which they define as 'the state whereby immigrants show, as a result of certain conditions, a setback in their assimilation process'.[27] The truth is that for many exiles the inner tensions are never resolved, and the result can be a progressively worsening alienation from the new land.

RETURNING TO NOTHING

In response to an assassination, the Czechoslovakian village of Lidice was reduced to rubble by the Nazis and most of the inhabitants were murdered. Nothing was left, and a resident who returned to the site many years later found that the 'most shocking blow' was 'to find just nothing there'.[28] Many others who have returned to sites they thought never to see again have been scarified by the experience. Naomi Rosh White's friend 'Renia' revisited Poland to be greeted, like Luka Prkan, with the accusation: 'You have survived. What for? Why did you come back? You have no right to be in Poland any more.' Another Polish exile returned to find that she felt nothing. 'It was not my country, not my house, not my shop.'[29]

Returning to the lost place, as Margaret Johnson found, can bring both pleasure and pain. A friend of Cam Trang Dinh is Do Thi Anh. In 1982 her four children and estranged husband escaped from Saigon without her knowledge. Desperate to find them and hating the communist government, she resolved to follow them. She told her mother of her intention to escape, but not the day or the route. After an alarming boat passage, she was placed in an Indonesian refugee camp and migrated to Australia under the humanitarian immigration program. She was reunited with her children, and in 1985 received custody of them.[30]

Like many other Vietnamese Australians, Do Thi Anh's passion for her Vietnamese homeland is undiminished.[31] *Australia ia a good place but Vietnam is home.* She misses the food, her relations, the colours of the paddy fields, her friends, the language, the skies. *Everything.* Tormented by homesickness, especially for her older immediate family, Do Thi Anh raised her children on Vietnamese traditional stories and histories. She pretended she knew little English, to encourage them to speak her native language. In 1993 Do Thi Anh had a job as an ethnic welfare worker, a busy social life and friends. When all her children were married she planned to live in Vietnam for half of each year. Would she come back to Australia if her children also returned to Vietnam? *Of course not.*

In 1991 Do Thi Anh returned to Saigon to visit her family. She chose simple clothes to avoid comment about her alleged wealth. She hated Saigon airport, so hot, so poor, so noisy; but reunion with her mother was wonderful, all tears and smiles. Her departure six weeks later was so painful that she asked her relatives not to come to the airport, but her eighty-seven year-old mother came part of the way on the back of a motorcycle to see Do Thi Anh for the last time.

Aboriginal children removed from their parents are exiles in their own land, and when they return to the sites of haunted memory, many of them journey to lay the ghosts of childhood cruelty. Bruce

Clayton-Brown, of Wiradjuri (southern central New South Wales) descent was eight years old and looking after his younger siblings when the welfare authorities arrived to remove his younger siblings from the care of his stepfather:

Dad's at work and I'm lookin after the three little girls, and the police said I was to ring Gosford police station. Dad came home with his little bag of shoppings, and I told him what happened, and looking back then I can see that's when he became heavily involved in the drinking, because his children were taken. The next thing I knew, three weeks later the police turned up to see why I hadn't been going to school, and they had a woman welfare worker with them ... Dad got frightened that he'd never see us kids again, so he packed us up and sent us to [a relative at] Bondi Junction. But the welfare tracked him down and finally said that if he didn't put us in St Joseph's Home at Kincumber [central coast New South Wales] they would take us and he'd never have any legal rights to access us kids again. So he put us on the train and we got off at Gosford station and we were met by the caretaker, Sister Eugene. A lot of this come back recently.[32]

There followed two years of institutionalisation which Bruce Clayton-Brown recalls with anger and affection, warmth and hatred. He respected Sister Eugenia because she was fair, *she made life bearable.* He knew that his five younger siblings were tolerably safe in the Home. He both appreciated and resented being *part of a machine,* that regularity and order in his life which he had not enjoyed before, nor for many years after, the Kincumber years. He objected to the compulsory church services and refused to attend or say grace at meal times. He would not eat rhubarb or 'cat's eyes' (sago) and was forced to sit at the empty table for hours. Resenting punishment, he would release the pigs or cows and was beaten with a strap or threatened with removal from the Home. He gave in: *All I could think of was that we had to stay together.* He learned the meaning of racism for the first time in Kincumber primary school: *We were the only Aborigines there and I had to fight nearly every day to look after my little brother and sisters.*

The anger and pain was the physical, emotional and sexual abuse he suffered in the Home and from some of the families to whom he was sent out on weekends and holidays. A certain nun would *play with us young boys when it was time for us to have showers.* Some of the temporary foster parents were worse. *Some were good, some were bad.* The worst crushed his spirit by bashing or by sexual and verbal abuse. One accused him, at nine years of age, of trying to rape two sixteen-year-old girls as he sat at the end of their bed talking to them.

Before he was eleven the Home closed, and Bruce Clayton-Brown's stepfather resumed the care of the children. After a hurtful row some months later Bruce left the family, and for a decade lived, drank and fought rough, sleeping on the streets of Melbourne, always on the run for a variety of offences or escapes, always on the alert for police or welfare officers.

Until his mid twenties Bruce Clayton-Brown refused to allow himself to meditate on his past. *I blanked it all out.* Then in 1987 he joined Link-Up, the association of Aboriginal people removed from their families who rely upon each other for counselling, support and guidance in returning to their own lost country. Listening to the histories of other exiles caused him to *start remembering the bad things as well as the good.*[33] *Hearing other stories. It started to eat at me. I finally let seventeen years of my life catch up.* Bruce signed himself into a mental hospital and found himself disbelieved by the psychiatrist to whom the experiences of exiled Aborigines were much more foreign than the concentration camps and ghettoes of the Second World War. Bruce's hatred extended to past authorities and relatives who had abused or rejected him. He was ready, he recalls, *to really do serious injury, to kill them and make the coppers kill me. I didn't realise how lonely I was.*

In 1995, towards the end of that long process of reliving and laying the ghosts of loneliness and pain, Bruce Clayon-Brown returned to St Joseph's:

I wanted to go back and get out a lot of memories out of there ... This is the first place I was sexually assaulted by females, and the first place I was sexually assaulted in the foster homes, and I blame them for that. All of them were churchgoers, so-called Christians, and I blame them the nuns for that, and I blame the welfare for putting me there in the first place.

St Joseph's (Brown Josephite) Home at Kincumber is now a retirement convent. When Bruce stopped his car outside the Home in 1995, he froze. This was no journey to nothing. *I got really scared. I couldn't figure out why.* There were the buildings, bigger, not smaller, than he imagined them. There was a sole remaining worker, who knew him: *Bruce! gee, you've put the weight on.* He found the visitors book and recognised names of former inmates, then met two men, also revisiting, who recognised him. He was pleased to learn that the authorities had tried to find him for a reunion. He *felt good after walking round.* He could not return to the rooms of darkest memory because they were occupied. Perhaps it was as well. In 1995 Bruce Clayton-Brown was a case worker and liaison officer for Link-Up. He has visited all the sites of his nightmarish childhood and drawn their capacity to hurt. Beyond the sites, though, trauma resides still. The ghosts of hurt and rejection linger beyond the sites where they occurred.

CHILDREN OF DEAD HOMELANDS

Cam Trang Dinh and Do Thi Anh remain in the new land because of their descendants. What of the children of such exiles, to whom the dead land, the half-grasped culture, the mysterious relatives, the abandoned sites and the stories—half myth and half history—are all that remain of their parents' obsession and enduring loss? Sometimes the refugees did not live long enough to pass on anything to their children:

> We are a lost generation.
> Our parents did not have time
> To rear us into manhood—
> They were killed too soon.
> Our children do not understand us—
> They grew up in a strange land.[34]

Judah Waten was bundled outside the room whenever the old Russian exiles gathered together:

> I was even afraid to put my ear into the keyhole. There seemed something mysterious and frightening about the conversation within. Rare words ocasionally did penetrate the door and wall. They were about births, and deaths, about deceased persons and the vanished past.[35]

The Dutch-born Cornelius Vleekens returned to Schiphol after an absence of thirty years: 'Took in my country. Took in my being. Took in my culture. Never to be lost. Never to be stolen from me again.' But Vleeskens was also 'More frightened, because I suddenly became aware that I am a stranger in Australia as well as in my native Holland, that I am no longer able to wholly fit in anywhere.'[36]

Sociologists have given much attention to the interaction between the first and second generations of migrant families. A prosaic view prevalent in the 1950s saw a 'static, monolithic local social structure around and within which there was an ongoing process of Australian identification into which individuals desirably and inevitably emerged'. A second model expected a new mainstream culture to be formed by the melting of all the older cultural elements. Later in the 1960s sociologists described individuals adopting 'a kind of cultural pluralism, mixing easily in some areas, keeping separate in others, and thus rejecting assimilation as the only solution to multi-ethnicity'.[37] Ruth Johnston, studying Polish refugees in Perth, observed that the first generation was taken as an unchanging reference group against which the children compared their observations of the outside world. 'The degree of anchorage in a specific culture seems to be the crux of the problem.'[38] This much is clear: the children most likely to suffer from painful divided loyalties are those whose parents retain close and unresolved emotional links with the old country. There are many such children living in Australia.

Senia Paseta was born in Australia to Croatian parents.[39] Her mother and father had escaped separately from a communist Yugoslavia that seemed, in the early 1950s, to have no future. They met in an Italian refugee camp, were separated, met again in Australia and settled in the Melbourne suburb of Brunswick. Senia Paseta was raised speaking Croatian as her first language. Her parents have remained Croatians first and last: *Naturally you hang on to your culture. You have nothing else. There was nothing here for them.*

When Senia Paseta was five the family returned to Croatia for the first time. Her parents had been absent for twelve years. Her clearest memories of the trip are her father drinking the seawater at Sali, his island birthplace, in ecstasy at his return, the warmth of her grandparents' welcome, of swimming for so long that she was sick. At the end of this first holiday the family very nearly remained permanently: they had made a sound start in Australia but in the early 1970s the Yugoslav economy was also improving.

In Australia the family continue to prosper. Senia Paseta's father put down economic and social roots. He made friends, established a house and garden, followed Australian politics, cricket and football. Her mother, working hard and maintaining interests in sport and politics, remained less well adjusted to the new country.

Ten years later, in June 1982, the family returned again to Croatia. In Senia Paseta's memory, the holiday was endless fun: partying, beaches, swimming, sleeping in. Her parents were less strict than they had been in Melbourne. She got to know the district, and the city of Dubrovnik: *beautiful, just so beautiful.* It was wonderful to do something as simple as leaving a film to be processed without having to spell her name to the assistant. Senia Paseta was beginning to know her relatives as people.

By September of that year the weather was turning colder. Croatians who returned to Sali each year from many parts of the world were preparing to depart. *Returning to Australia was the most awful day of my life.* The whole family, it seemed, cried all the way back to Melbourne. Senia Paseta remembers that she took months to recover. She refused to speak English, listened to Croatian music and found it difficult to relate to her non-Croatian friends. *Where you have to go back to is not necessarily home.* She had wondered about staying behind. But where would she stay after her parents left? Where would she go to university? For the first time she noted a curious fact: her father, visiting his native land, did not seem to mind leaving it as much as she, a native-born Australian. The following July, when she knew that her people would again be gathering at Sali, 'doma'—the homeland—was always in her thoughts. At eighteen, after leaving school, she returned to Croatia for nine months. Australia seemed further away than ever, the home of friends rather than a place of deep attachment.

Yet Croatia was becoming more complex. Sali was beautiful and wonderful for a holiday, but as an adult Senia Paseta recognised it also as small, parochial and gossipy. Among her relatives was a discernible lack of interest in Australia. Her mother had been one of seven children; her grandmother was so busy that her most obvious emotion was not curiosity about Australia but pleasure that her emigrant daughter had done well. There were tensions between different sides of her family, manifested by the decision about which relatives to stay with. Senia was conscious of resentment directed at her money, as if wealth was easy to come by in Australia. *It was awful watching the women stand back to let the men eat first.* She realised that rather more of the intellectual and social life which had made her was back in Australia.

When Senia Paseta returned again to Croatia in 1992 the civil war had been in progress for over a year. Zadar, one of the cities she loved most, was almost in ruins. The Second World War bullet scars in the walls which her father had shown her had vanished along with some of the buildings. A friend had been killed. *This is me being torn apart. It's never going to be the same again. I was so distressed that I nearly had a breakdown.* Worst of all was resentment by some Croatians at her presence. Australians weren't suffering. *What would you know? You've not been here* was the implied question: her own countryfolk could not understand her anguish, that she wished she was suffering as they were.

When the time came to leave again, some relatives had become very close but others were uninterested and did not bother to see her off. *In a way it didn't matter, because Croatia was my spiritual and emotional home.* She wondered about remaining, perhaps to teach English. She told a relative *I want to stay there, this [country] is my home.*

Leaving was terrible. Because we had to leave them in the middle of this war, and we knew we could leave. It just broke our hearts. Shocking, it was just shocking leaving. To get to Dad's island we get a ferry from the mainland, and that ferry, the last ferry trip out is always just hell. It's just hell. Always tears, the last view of the island, and it's always just six in the morning so that the sun's coming up and it looks magnificent, and everyone's always standing at the port. Horrific, it's just horrific, and last year was worse because it was so dismal. And there was nobody there, nobody came. It was dead. You could just hear bombs all the time.

Meanwhile her parents were feeling uncertain about their own future. If Senia and her sister chose to live in Croatia, she believed their parents would follow them. Her mother might stay for good, her father might return periodically to Australia. The parents welcomed, in a sense, the indecision of their children as, in addition to other factors, they wanted to be with them. In 1994, when Senia Paseta was in her mid twenties, belonging anywhere was a matter becoming steadily more complex. She would want her children to know their Croatian language, their cousins, the magical coastline, but they would need to know Australia also. *It's such a huge part of me, it's made me in lots of ways.* She felt affection for Canberra as well as Melbourne, and a deepening love for Dublin, where she was carrying out post-graduate work. *I make it worse for myself by having new places.* She planned the next phase of her life in Europe, if not in Croatia itself. *The Dalmatian coast. And that village. That's my place.*

Con Boekel was born in Indonesia to Dutch parents and spent his first years there.[40] Shortly after the Indonesian war of independence his father migrated to Australia. At the age of seven, he accompanied his mother on a trip to Holland, where his sister was born. Six months later the family returned to Toolangi near Healesville, Victoria. They lived and worked in shops and farms about Melbourne and eventually bought a dairy farm at Koo-wee-rup. Con Boekel received his secondary education at a Catholic boarding school. Always he felt restless, and at the age of forty he had stayed in any one place no longer than four years. His parents had apparently resolved their own ambiguity because they knew the physical realities of that from which they had exiled themselves, but where, he asked himself, did he really belong? A reconnection with Indonesia he thought to be problematic: he has not revisited that country because of possible local resentment of the Dutch, because he no longer had any family connections there, and because the version of history he learned from his parents was so different from the version taught in Australian schools. *The traditional heritage of a place depends on people believing the same relative notion of what the truth was. I was conscious that I was not sharing it with others, and that other truths were possible.* Where he belonged, and where he belongs now, is one of the central uncertainties of his life.

Because his parents moved so many times, Con Boekel feels no particular connection to any place in Australia. Even the thirty-year residence at Koo-wee-rup brought little stability for, as his father aged, the farm diminished:

They sold most of the farm. And so one of the things I'm conscious of, [is that] when I was a kid there was actually more of the area that I could move freely through than I can now. When I was a kid you could walk 3 or 4 k [km] across people's property and walk along the shores of Westernport Bay ... If I did it now as an adult I'd feel ... as if I didn't have that freedom. Because the farm was sold it just reinforced my notion that all this was ephemeral, not fixed. If one of my brothers had stayed on the farm, as they had in Holland, where they'd been on the same farm there for centuries and centuries, literally, then it would be worth investing ... but for me clearly, mum and dad are on a 20 acre block, they've sold bits and pieces of the farm off to live off in their retirement, and when they die none of my brothers and sisters will probably live on that place. Even though it was quite an important place for me as a kid, it's like shifting sand. It just reinforces my notion that it's a wasted effort. When dad sold [the first block] it was gone. The person who bought it ... was training racehorses, he got rid of the dam and bulldozed the trees. It was just like a razing of all those wonderful areas. The notion of ephemeral, that places were ephemeral.

It is the older children of Peter and Betty Boekel's family for whom Holland is mysterious. The central ambiguity of Con Boekel's life lies beyond the lost acres of the farm to the birth country of his parents, not lost to them but lost to him. Though he has spent almost all his life in Australia, Con Boekel has never felt at home anywhere. His parents spoke mostly English to the children, but they talked about Holland. He heard news about his Dutch relatives, but feels disconnected from them. He knows the family history intimately, but second-hand, from books and conversations. He knows about individual local characters, the robust regional humour, the village fair, the local dialect, the Nazi invasion, the wildlife, even the special sheds in which cabbages were stored. There is the place, the country, the land—but where is the continuity to link him with that immense and historic past? It seems to stop, abruptly, with his parents. What is at the heart of the familiar, yet ultimately inaccessible, family history centred for so many centuries in a couple of small Dutch villages? Where is that meaning?

Both my grandfathers lived in the same house for centuries. If one of my ancestors from the seventeenth or eighteenth centuries had come to that farmhouse in about 1916, when my father was born, he would have felt pretty comfortable. The horses would've been doing the same thing, they would've been going down to the same paddocks, they would've been growing the same crops, they would've been using the same agricultural techniques, they would have been picking things into wicker baskets ... I think that the sort of place you get from that sort of continuity, generation after generation after generation, things don't really change. Immutable.

The result of that abrupt discontinuity in 1952 is the placelessness of the child of the exile. Con Boekel feels that the absence of physical roots has made him more self-reliant. Placelessness can liberate: though in some ways envious of Australians who can trace their family history over 100 years to the same valley, he feels that attachments can be a physical and emotional trap. *I would never die for a place. I'd never join an army and die for a place, and I don't comprehend it at all.*

Whence arose this profound and ambiguous rootlessness? Was it the years at boarding school, or the contrast between standard British history and the Dutch accounts which held the British colonials to be 'imperial bastards'? Was it the restless shifting of his parents in their first Australian years, or the harsh severance from an ancient heritage which the parents imposed on their children? It was no accident, he thought, that the parents had taken none of the children to Holland on any of several later visits. Con Boekel is aware that his parents did not want him to return to Holland to marry, which would have implied that their own major life decision to migrate had been wrong. He feels Australian in his love of the bush, gum trees, beaches. To step out of a plane and *feel the tropical warmth bathe me was a moving experience.* He appreciates personal freedom, the decencies of Australian everyday life, the openness. Is he, then, Australian or Dutch?

'Are you a Dutchman?' That's an interesting question. I don't really know, and that's a central uncertainty in my life. Part of the answer is—to what extent am I an Australian? In some senses I'm neither and in some senses I'm both. We went back to Holland when I was about six or seven, and one of the things I did when I was there was, grandfather (my mother's father) took me fishing. I remember going on a boat in the canals, picking potatoes into a wicker basket, going to school there, the little canal behind my mother's home in Holland, my relatives, the only time I saw some of them ... My cousins are living in the house my father was born in, so it's nice to think that that's still happening, it's been like that since 1700 ... One of the sorrows in my life is missing that sense of connection, being divorced from that sense of connection

and knowing that's gone. So I suppose to the extent that's a sorrow, I have to be Dutch.

Australian sociologists were right to focus on the relationship between the migrating generation and their children, but their research has fixed too narrowly on school achievement and adolescent identifications. Allegiances grow more complex in maturity. Planned or not, the parents of Senia Paseta bequeathed to their children love for the lost country, the Boekels bequeathed ambiguity. Susan Karas imparted rejection and Cam Trang Dinh imparted sorrow.

What is to be transmitted by those in mourning for dead homelands? Elie Wiesel, the Jewish intellectual, concluded 'I will tell my son that all the fires, all the pain, will be meaningless, if he will not transmit our story to others, to his friends, and one day to his own children ... What distinguishes the Jew is his memory ... If we stop remembering we stop being.'[41]

Some 7000 Jewish Holocaust survivors, mostly from Czechoslovakia, Poland and Hungary, came to Australia before 1950. Their past is ambiguity and chaos. So are their memories and so is the history passed to their descendants. Elizabeth Wynnhausen, whose Jewish Dutch family migrated as displaced persons to Sydney, listened in silence to her parents' fragile mood and hectic gaiety as they reminisced about the past. She sensed the desolation the words concealed.[42] Theirs was a communication of history beyond speech.

Fragments were also the experience of the Victorian writer Arnold Zable. His family came from Bialystok, on the Russian–Polish border, where almost all the previous generation of his kinfolk had perished at the hands of the Nazis. The Jewish elders in Melbourne were passionately anxious to communicate the meaning of lost pasts and dead sites to their descendants, but what was that meaning? To Zable, their broken messages were like the 'Siberian night sky, streaking starlight between spaces of darkness', never complete but dissolved 'into an infinite darkness they called the Annihilation'. Such fragments of history he named the ashes and jewels of his, and his family's, and the whole of the European Jewish community's, lost and dead homelands.[43] Arnold Zable's Europe:

> is littered with reminders: stories, plaques, monuments. And beyond the physical borders, the echoes of what happened just one generation

> ago, on this soil, reverberate in the dreams of survivors; and the children of the survivors. They also have been drawn into this landscape of darkness with its aborted stories and its collective memory of suffering.[44]

The Australian children intuit these many meanings, but in the end Europe is their parents' lost country, not theirs. Zable concludes:

> We [Australian Jews] were born into the wake of the Annihilation. We were children of dreams and shadows, yet raised in the vast spaces of the New World. We roamed the streets of our migrant neighbourhoods freely. We lived on coastlines and played with new horizons. Our world was far removed from the sinister events that had engulfed our elders ... 'You cannot imagine what it was like', our elders insisted, 'You were not there.' Their messages were always ambiguous, tinged with menace, double edged: 'You cannot understand, yet you must. You should not delve too deeply, yet you should. But even if you do, my child, you will never understand. You were not there.'[45]

In search of the meaning of the fragments, Zable caught the train from Warsaw to Bialystok. He felt the past expressed in those physical sites, so familiar yet so strange, closing on him more tightly than ever it could in Australia.

> Word associations emerge and impose themselves on the countryside rushing towards me—exile, prison camps, pogrom, interrogation: fragments of family legend and communal remembrances. It is an ancient fear, handed down through many generations, lying dormant and liable to be triggered off unexpectedly.

Zable found an ancient forgotten cemetery where his ancestors lay buried. He read a note attached to one of the ovens in the sole remaining crematorium at Auschwitz: 'We promise to show our children where our grandparents hugged for the last time.'[46] He stood at the site of the Great Synagogue of Bialystok in which 1500 Jews had been burned to ashes by the Nazis. Here was revelation. This was no longer the site of the 'mythical landscape, in a remote kingdom of Darkness, in which ancestral ghosts stalked unredeemed.'[47] The past was real because the site was real; the meaning of the fragments lay in the landscape itself. The chilling but wordless past, the intense, passionate but broken narratives were made whole in the human site and field of action, that place.

Zable now knew that the good and evil of lost and semi-mythical homelands do not exist untrammelled by physicality. They have exact locations. Standing at the site of the Great Synagogue he understood that each site acquires the characteristics of the human actions perpetrated there and, conversely, that the deeds which belong to the past lose their meaning if they are denied the real sites where they occurred. Susan Karas's past is real because her country is real. She was there. Arnold Zable was not there, and the Jewish past became real only when the imagined place became real. Until such moments, the lost places of the exiles are not quite those of their children. In the 1980s there were hundreds of such Australians in search of the lost homelands, 'wandering country roads, and city streets, picking our way through forest undergrowth to uncover mould-encrusted tombstones ... Perhaps this is how it has always been for descendants of lost families.'[48]

We are not yet at the end. If the lost country is revealed to the exiles' children only at the site of that loss, at what sites does the new country become real? Andrew Riemer found that there was no such place.

Riemer was born in Hungary in 1936. He narrowly survived the anti-Semitic purges of war-time Budapest and migrated with his parents to Australia in 1947. He had an orthodox education: high school and the University of Sydney where he became a lecturer in English. Like Con Boekel, Riemer knew his family history so well that at one level Budapest was a city more intimately known than Sydney, but as he approached his fifties he was increasingly haunted by the dim shapes, images and colours of long-left Budapest. He had a childhood memory of a laneway at dusk: 'I experience a sense of great peace and contentment, mixed with an acute sense of loss.' In 1991 he returned to his birthplace for the first time since the family left on a dark and bitter afternoon over forty years earlier.[49]

Here was that first ambiguity familiar to Arnold Zable. Budapest, where Riemer felt he belonged more intimately than in the city where he grew up, was also the site where most of his family was killed or started their journey to death. This was the 'life that had been poisoned for us by hatred and barbarity, yet a life we yearned after in our exile'. He visited the Jewish museum: his perceptions were perhaps fresher for having been away. He found one of the homes which his family had occupied in the last months of the war. He chatted to its occupant. The discussion began cordially but ended sourly when the occupant misinterpreted Riemer's interest and nostalgia as a desire to reclaim the house. What had the experience of returning to the lost country taught Riemer? That he belonged emotionally neither in Budapest, that city of terrible memory, nor in Sydney, the familiar place which he called home.[50]

For the first time Andrew Riemer realised that his alienation from both Sydney and Budapest was more than emotional. The realisation came to him one afternoon when, as a guest lecturer sponsored by the Australia Council, he was explaining some Hungarian history to a group of interested and sympathetic Australian tourists. They listened, and were silent before the magnitude of the narrative. But Riemer sensed that the atrocities of Jewish Hungary, so personal and sensitive to himself, were to the visitors merely parts of that universal human predicament which included South Africa, Nicaragua, Chile and East Timor. The native-born Australians cared, they were grieved, but no more than that. The passion of these apparently decent, compassionate Australians was reserved for events closer to home. At last Riemer understood that it was not only because he looked different from other Australians that he felt alienated; it was not only because he held a traumatic and collective history not shared by other Sydneysiders: he, the outsider, was not fully in sympathy with their concerns either. Of the Aborigines he reflected:

> I cannot participate in these [Australian] people's sense of personal or national guilt because, for me, the terrible persecution and exploitation of the Aboriginal people of Australia occurred, at least in large part, at a time when Australia had not existed for my family ... much as I understand the terrible perplexity which this consciousness does and should impose upon those people, for me it is not something that has entered the fibre of my being ... It is not part of my history, it is merely a part of my recognition of the terrible cruelty of mankind. For this reason there will always be something of a gulf separating us. I am always conscious of being outside the emotional currents of Australian society.[51]

Is Susan Karas completely at home in Sydney, and Irina Petrovna in Canberra? Will Cam Trang Dinh and Do Thi Anh be content in their own countries if they return to stay? When does 'them' become 'us', 'I' become 'we', 'these people' become 'my people?' Is there, for some, no country to which they can return—or advance? Dead countries are the most difficult of all dead sites to negotiate.

CHAPTER 3 NAMADGI: SHARING THE HIGH COUNTRY

Granville Crawford's country was a corner of south-eastern New South Wales. It ran 100 km from east to west, from Michelago to the Murrumbidgee headwaters near Coolamon, and north–south from Yaouk to Tharwa. After a residence of fifty-seven years Granville Crawford knew that land *as well as the average bloke knows his backyard.*[1] Yet his deep attachment to this land was shared by others. The Ngunnawal Aboriginal people, bushwalkers and environmentalists knew and loved the same valleys and ranges. Now this land is not only emotionally loved but emotionally contested.

Granville Crawford knew his country almost from birth. He was born in Queanbeyan, near Canberra, in 1929; his father died in an accident just before he was born and he was reared at Naas, a tiny settlement in the foothills of the Brindabella ranges. His foster parents were his grandmother Bertha Dyball, who won the contract to bring mail to the remote stations in the hills above Tharwa, and his step-grandfather Herbert Oldfield, a high-country sheep and cattle farmer.

In a world where the sexes lived their lives more separately than today, the old bushman had the greatest influence on the boy. Herbert Oldfield showed Granville Crawford his own special places: *That's where Hi [I] used to turn the cattle,* a sparkle in his eyes. Once Oldfield gave him the last of the tucker because a young feller like Granville needed it more than an old feller like him. *I could show you the exact spot where he was sitting. I can still hear the rattle on the slab floor when he came on to the verandah of that old hut. You'd always know who it was by the creaking of the slabs.* In the alpine grasses of the Snowy Mountains high plains, Herbert Oldfield taught the boy how to care for the rock wallabies, and how to stop the stock damaging the more delicate vegetation. *Don't ringbark that tree, there's not many of them.* Bushmen won respect by their accuracy in predicting snow, rain and frost, their skill as riders, their respect for and knowledge of the country and the ways of stock. Other bushmen like A.W. Bootes and Ted Brayshaw had a lasting effect. Bootes told Granville to *keep your horse on the trot, don't canter or it'll be the signal for your cattle to panic.* Oldfield imparted an aphorism which Granville Crawford had reason to remember much later in life: *You can never take an old brumby away from where he was reared. He'll never ever do any good.*

On Saturdays from the age of nine or ten Granville Crawford would travel by horseback, on behalf of his grandmother, up the road from Tharwa to return by the back country, visiting each mountain station to pick up the mail. From this early age the stillness of the bush began to enter his soul. During the manpower shortages of the Second World War he drove and worked the stock alone. Granville Crawford learned to ride at six, rode alone at nine, shore a sheep at twelve, broke his first horse at fifteen. By then the different seasons and the ever-changing weather had embraced him. *Above all for everyone who's lived in that country and loved it, is that feeling of peace. You feel safe, you feel secure, everything's familiar, you're in control, you know the seasons. You've seen it all before.* The mountains shaped the character of the men and women who lived there. His grandmother and Herbert Oldfield, riding miles for their supplies, fetching water from the creek in buckets every day—*they knew the real pain of poverty.* There were vivid memories of a roaring fire in a mountain hut while the wind howled and the rain came down in torrents. Would he have loved that hut so much, he wondered, would the memory have been so vivid if he had not spent the day in the open, and the previous half-hour feeding and tending his horse and dogs in the pouring rain? Attachments formed out of memories and memories formed out of experience. What attachment, what memories? Granville Crawford's country was a single entity: mountain ash, blue haze, snowdrift, deep shade, chill breeze in midsummer. Simultaneously it was a vast particularity of known sites, finite events, exact times: this man at that place.

Such attachments were intimate and personal, but Granville Crawford's land was also socialised. In 1952 he married Rae Gregory, who ignored the advice *that little feller's no good for anything off his horse.* Her family too had been in the district for over a century. For thirty-five years Rae and Granville raised their four children and lived at stations at Glencoe, Orroral, Gudgenby, Honeysuckle and Naas. Homesteads were memories of the children born in them, of relatives who stayed, of the schools, the horses, the buildings. *We always lived within hearing of a running stream, and at Gudgenby you could predict the degree of frost by the sound of the water coming down the gorge. I could take you to the exact spot.* In the socialised landscapes Granville Crawford still drew on the simple power of close observation: while living at Glencoe Station he noticed how the magpies would start the day on the eastern side of the creek and follow the sun to the western side by the end of the day. There were special places away from the homesteads: that long-remembered spot where he buried his best dog, the yard where he mastered a particularly high-spirited horse, the well where he kept trout alive for a few days after catching them in the creek. *I could show you the exact spot.*

Perhaps because he spent more of his life alone, Granville Crawford found his ultimate fulfilment not in family life, not in the valleys, but up in the ranges in solitude. Horse and dog made a team. *Old bushies talk to them all the time. They make great conversation.*

And there's nothing more beautiful than to sit on your horse on the top of a high mountain and just look, just look about as far as you can see. There's mountains rolling away, gullies and gorges and all the different colours according to the angle of the sun, the time of the year and the temperature.

Ultimately the mountains meant peace and personal fulfilment.

A bloke's not blabbing in the pub about what he could do, it's just something he experienced himself and you share the joy with yourself. There's some very special spots, very personal special spots, and I guess that everybody ... who lives in open spaces has some very special places where they feel security. I don't wish to go into a great spiel about faith and all that sort of thing, but there's ... some very special places where you can feel very near. And that's a great feeling.

Granville Crawford was a sheep farmer—cattle were too expensive—but the routines of cattle and sheep farmers in the high pastures were not dissimilar. The revered points of the year were the great droves to and from the high alpine grasses. A week before the end of November each year Granville began to prepare the gear and rations needed for the ascent. With a mate, a packhorse with the rations and two or three dogs he would begin the trek from the valley floor, from Glencoe Station. The first night stop was Gudgenby, Bradley's Creek the second, then a short day to Yaouk on the third, Yaouk to The Circuits on the fourth, and a short day to the lease on the fifth—his own snow lease, the paddock known unromantically as D10. There they'd camp for a few days at one of the now famous huts of the high country, usually Crace's Hut, riding the boundaries, mending the fences. Often they would stay for a few extra days exulting in the air, the serenity and the space. Then Granville Crawford, mate, horses and dogs would return home, back to Glencoe Station to grow a summer lucerne crop or two. The sheep would remain on the high plains until the last day of May, when by law they were required to be droved out of the alpine pastures to spend the winter in the valleys 1000 m below.

This high plains country of the Snowy Mountains National Park is shown on modern maps as the Jagungal and Bogong Peaks wilderness areas, not far from the Talbingo Dam, marked as 'Winter: road closed' and 'Difficult in wet conditions. Unsuitable for 2WD'. Though all that land is now inaccessible to pastoralists, Granville Crawford's face lights up at each name. The Cotter: *Sometimes we'd kill a sheep and salt it.* The Pockets: *I put in a lotta time there after the water race at Murray's collapsed.* Crace's Hut: *that's gone now.* The Pockets to Coolamine: *That's about a day's sheep walk.* Oldfield's Hut: *That's where I used to go when I was a kid. Still there—a big place, stable and a feed room.* Coolamine homestead: *I took the sheep there for Roy Tong once or twice. I'd give anything to see the old man ride up, and as I get older it gets more and more intense.*

These oddly named places like Oldfield's and The Pockets are mountain huts and former homesteads. Most were built in the first decades of the twentieth century, and now are as well known to skiers and bushwalkers as they were to the pastoralists. The alpine huts were built to provide shelter from westerly winds, to catch the morning sun, stand close to wood and water and look out on to grassy plains. Oldfield's, where Granville Crawford has spent many a night, has a *very long verandah, slab walls, rusty iron chimney, verandah posts, slab floor and a holed water tank.* It possesses distinguishing features—'Please

Knock' printed on a wallaby skull and, by the door, the hand-written slogan 'Wombats of the World Unite'.[2]

At this point pastoralists like Granville Crawford enter the mythology of southern Australian folklore. Their landscapes, as D.J. Mulvaney put it, are 'deeply ingrained in the Australian psyche'.[3] The image of mountain men and women, complete with dogs, Akubra hat, Drizabone riding coat and elastic-sided boots is still almost as familiar an icon to most Australians as when Banjo Paterson wrote 'The Man from Snowy River'. The pastoralists seem to find it easier than most to articulate their attachments to the land as 'sacred' or 'mystical'. Leona Lovell, a Victorian cattle farmer, echoed sentiments as intense as Granville Crawford's: 'The mountains are almost like our church—that's where we go to fulfil ourselves so that we can carry out our lives. We've done it for generations.'[4]

Alpine graziers felt a special sense of belonging to the high pastures. Bill Hicks, a Victorian high plainsman, described the annual return to the alpine pastures on the night of a December full moon: 'Your heart pumps a bit faster, and you feel excited and peaceful at the same time. The horses know where they're going and the dogs too ... the people and the livestock ... everyone ... There is this feeling of heading to where you belong'.[5] The pastoralists:

> claimed a special affinity with the land, a sense of belonging, a hereditary responsibility to take care of it, an atavistic, personal identification with the High Country that has never been sought and has never been cultivated. It is bred in the bone ... They all have it, and they cannot change, because it is part of what they are ... mountain cattlemen.[6]

Their qualities of toughness and pioneering have been compared by admirers to those of soldiers of the First World War:

> Young men in caped coats and broad brimmed hats tending their horses, discussing quietly the serious things of the day. And suddenly I saw what they really were—the living ghosts of splendid young men who 70 years ago stood side by side, looking up the parapet of Gallipoli.[7]

Granville Crawford made his first ride to the high plains when he was in short pants in the late 1930s, when the alpine droves were at their peak. He drove his last mob of sheep in 1965 when the snow leases, were, one by one, being closed. He felt and articulated a sense of belonging to certain sites and landscapes equal to that of any other non-Aborigine. The high country pastoralists had impressed most other Australians that their intense attachments to the alpine pastures were not only real but legitimate. From the 1970s both that reality and its legitimacy were challenged. The high plains, the ranges and the valley farms, for different reasons, were soon to become Granville Crawford's lost country.

In the 1950s there were several uneasy currents tugging at the pastoralists of the Australian Capital Territory. A campaign to remove the stock from snow leases began with the declaration of the Kosciusko State Park in 1944, and opponents of summer droving continued to allege the destruction of rare species of grassland and wildlife by sheep and cattle. The Snowy Mountains irrigation and electricity engineers' desire to protect the water catchment areas above 4500 ft (1370 m) added to the campaign to put an end to the annual pilgrimage to the snow leases. The principal argument against the mountain pastoralists was scientific—that the cattle and sheep eroded the soil, denuded the vegetation and destroyed rare plants, and that the degradation of the whole of the Bogong High Plains water catchment in Victoria was probably caused by overgrazing in the 1940s. The soils of the mountain grazing country are friable and soon eroded when the covering vegetation is removed by fire or overgrazing.[8] During the 1960s the snow leases began to be withdrawn; by the end of the 1980s, all the New South Wales and Victorian high country leases had gone or were to be phased out.

Meanwhile a second and more local campaign began, to drive the farmers out of the southern valleys and hills of the Australian Capital Territory. All their lives Granville and Rae Crawford knew that the federal government would at some point resume the farming country to the south of Canberra. It might be needed for the expansion of the city, for a dam, for a national park, but the land must—in their lifetime or their children's—be resumed. *We never dreamed it would come.* Though some farmers left under duress during the 1940s, the major effort to create a national park in the Australian Capital Territory was opened in 1963 by conservationists and bushwalkers.[9] A year later the Orroral grazing property, which Granville Crawford had once managed, was resumed by the Commonwealth government for the American orbital tracking station. Although a national park was not proclaimed until 1979, steady pressure was put upon graziers throughout the 1970s to quit not only their high country leases but the whole of the southern grazing leases of the Australian Capital Territory. In 1968 Rae Crawford read in the *Canberra Times* that the twenty-nine property owners in the Tharwa area would be given three months to 'treat', that is, to negotiate the removal of themselves, their stock and belongings from the valley. There would be free agistment of sheep and cattle for the first year, leased agistment in the second, then the government would resume the land.

The Orroral valley farmers had become caught in an ideological battle over whether a national park should present human or wilderness values. Grazing, not just on high country but in the valley lowlands, had become unwelcome to many supporters of the wilderness which supposedly replicated the wild country predating white settlement. Bob Brown, the well-known conservationist, declared contentiously that 'wilderness is a large tract of entirely natural country'.[10] Wild country was beginning to signify a rejection of any human traces in a 'natural' area:[11] in 1988 the president of the National Parks Association asked why the Orroral graziers should be allowed to remain once Gudgenby (later extended to the much larger Namadji) became a national park. It was only cultural cringe, he argued, which made it seem as if wild areas needed 'the human hand to make them complete'. There was plenty of other grazing land nearby. The wildlife was in danger, but 'our ubiquitous grazing landscapes' appeared 'little threatened'.[12] Another opponent of the valley farmers claimed that grazing reduced the diversity of species and caused erosion, siltation and pollution.[13]

In reality there was no doubt that the forests and valleys of the Brindabella ranges, like other areas of southern Australia, were an essentially human-created landscape. Besides the farmers or pastoralists, Aborigines had over several thousand years created the landscape into which Granville Crawford was to set his roots. D.J. Mulvaney responded that Gudgenby—in effect, Granville and Rae Crawford's country—was an example of responsible grazing management. The area was no longer a fragile wilderness ecosystem, but had been farmed continuously since the 1820s. Grazing should be preserved because it demonstrated 'historical inter-relationships, continuity over time and strong historical associative values'.[14] A supporter of the farmers wrote, 'I for one would prefer a herd of cattle on Gudgenby than a herd of pigs in the [homestead], and that is the choice we are faced with'.[15] Despite a belated formal acknowledgment that the cultural resources of the area were of equal importance to the natural resources, the Gudgenby National Park was proclaimed in April 1979.[16] By then most of the historic buildings had been vandalised, destroyed or demolished, some evidently in the wider belief that 'if you say you should retain the cattlemen's huts you are saying in effect that you approve of grazing'.[17] The lost homesteads included that built on the property first occupied by Alexander Crawford in 1844.

The values of untouched wilderness prevailed and all the valley leases were ordered to be terminated by June 1989. The Australian Capital Territory Parks and Conservation Service planners affirmed that grazing was 'inconsistent with current Australian national park philosophy'.[18] In the course of planning, the Orroral Valley—Granville Crawford's first country of memory and attachment—was designated 'low-level bushland recreation with provision for public access for bushwalking, orienteering, camping, public appreciation and education'.[19] The National Capital Development Commission planners recommended that the eastern portion of Gudgenby station be kept as open grassland, without discussing how, in the absence of grazing animals, this might be achieved. The western side was to be allowed to regenerate naturally.

The invitation to treat and the certainty of ultimate eviction caused a creeping decline in the morale and activities of the valley farmers. For the next two decades, the multitude of government departments involved in the creation of the park dithered about policy. Fourteen separate statutes were in force in the park's management. For a time the farmers were allowed to remain on a weekly basis only. Improvements were pointless. Employees changed so frequently both in the field and in head office that nobody, in Granville Crawford's recollection, seemed aware of overall direction or planning. Officials seemed ignorant of the basic principles by which the old bushmen had survived. A ranger told him that a fence would be put along Honeysuckle Creek and that his stock must stay outside it. How would they drink in a dry spell? A few months later another ranger told him that the cattle could have the creek but would be debarred from the upper pastures. Where would the cattle go in the next big flood? Didn't the authorities know, Granville Crawford asked himself, that cattle ruined flooded country if they were forced to remain on it? Faced with declining income and narrowing prospects, the Crawfords concluded that they should accept the resettlement allowance and quit the valley.

It was 5 March 1986. Rae Crawford felt a sense of relief from the *long, dark tunnel* which would continue as long as they remained in the valley. Here was no future for themselves or their children; the terms of resettlement elsewhere were generous. Less bonded with the mountains, or too busy, she felt no sadness. *We'd had lots of moves before, it was an adventure, and we hoped to benefit—and we did.* Granville Crawford's memories are much more painful: *I'll never forget it. Bloody awful. All the sheep, cattle, horses, the dogs, all our furniture, all our belongings, the whole lot—went.* The last fifteen sheep were loaded into the trailer of the ute. A friend, Pat Reid, *came out in his horse trailer and we loaded my horse. It was the last animal to leave the place.* In his diary Granville Crawford wrote, *I'll be glad when we are installed in our new home at Sherwood Park.*

Sherwood Park was a 366 ha property on the south central plains of New South Wales. It was productive country which in 1994 Granville Crawford felt that he could respect but not love: *I doubt if it'll ever be that real home here, but it will be to Rae.* He has not tried to reset his roots into the plains; belonging in new country *has to come from within.* In the first years after the move, he felt totally alienated. One by one he sold his horses: there were no hills to ride on, so what was the point of keeping them?

Sherwood Park adjoins the abandoned village of Junee Reefs, whose inhabitants left many years ago after a minor gold rush. A recreation hall and a war memorial, both now kept neat by Granville Crawford, are all that remains of the school, hotel, post office, tennis courts, stores and park which once comprised the village. To the outsider it is a pleasant though rather melancholy abandoned hamlet in the gentle hills of the Illabo district which speak of prosperity, care and security. To Granville Crawford the rolling treeless undulations, the heat, the dust, the ordered fence-lines and the red-brown earth are strange and alienating. The year he arrived at Junee Reefs, he wrote a poem about the experience.

Really I was due back in Canberra to chair a NSW farmers meeting, and there was a foul up with the trucks. They didn't arrive until late. I was sitting out there on the front verandah and the wind was blowing across the stubble paddock with these great swirls of dust. It was dry and it was desolate and it was hot. And I thought to meself, 'My God, what have I done?'

The First Summer at Sherwood Park

As I stand out on this open plain
I think of the place from where I came
Where the everlasting daisies grow.
That's where I would like to go.

Where the alpine ash grows straight and tall
And the lyrebird to his mate he calls
And the magpie with his joyous singing
Sets the mountain ranges ringing.

For the mountain air is so pure and clear
That every bird call one can hear
Now the red dust blows about my feet
And my body wilts from this dreadful heat

What I would give once more to roam
In my beloved mountain home.

Granville and Rae Crawford occasionally revisit their country of memories. They are bitter about its rapid degradation. Orroral homestead is gone, so is their cottage at Naas. Soon after they and most of the other Orroral families left, pigs, wild dogs, rabbits and noxious weeds began to multiply. Fences fell down, houses lay empty and were demolished or vandalised. Gardens were overrun and pastures overgrown. Granville Crawford took his son James to have a last look at Horse Gully. *It won't be the same again.*

PASTORALISTS AND THE WILDERNESS

How have Australians judged the attachments of the valley farmers and the high plains pastoralists to the country to which they thought to belong forever? Canberrans who loved the country for its wildernesss values found it hard to share their attachments with farmers who had used it so differently. An article in the *Sydney Morning Herald* extolling the Namadgi National Park to bushwalkers stated that the farming country now was empty because the farmers had failed and abandoned their holdings.[20] Heritage surveys in the 1970s of what later became the Namadgi National Park reported a number of traces of Aboriginal occupation, but that 'There are no buildings in the area which, in our opinion, justify preservation, except the slab building near Bobeyan.'[21] Yet the National Capital Development Commission draft Policy and Development Plan had found at least three ruined homesteads, as well as many other gravesites, bridle and bullock tracks, stockyards and fences.[22] Nor was the Namadgi assessment the only example of belittling the achievements of the settlers. The Australian National Parks and Wildlife Service 'removed all structures not fit for living'[23] from the village of Kiandra—which meant nearly all of them. It was as if the old farming and grazing areas were to be forcibly expunged from human memory. The legitimacy of the farmers was challenged and, as if in necessary consequence, the attachment of the farmers to their land was discounted or ignored. The National Parks and Wildlife Service predicted that 'it should be possible to arrange alternative accommodation for those [Orroral valley] residents without undue hardship or great expense'.[24]

The hostility towards grazing within the Namadgi National Park, to which Granville Crawford fell victim, emanated in part from a concept of wilderness which had its clearest expression in the United States from the 1850s, which held that spirit resided in the waste places of the world 'unworn by man'.[25] It also shaped the debate about the high country pastoralists. No doubt it was true that the stock had harmed the country, especially in the nineteenth century, but the pastoralists replied that outsiders with T-bars, four-wheel drive vehicles, ski huts and roads had also degraded the country, and that, whatever the damage wrought, the grassland vegetation had reached equilibrium under existing moderate grazing pressure. Victorian cattlemen and women marched through Melbourne streets bearing the slogans 'Cattlemen are true conservationists' and 'Alpine grazing reduces blazing'.[26] The historian Tom Griffiths remarked that 'although their scientific arguments have been refuted or doubted, their heritage arguments have rarely been addressed'.[27] Griffiths was right: the feelings of grief over lost country experienced by people like Granville Crawford were set aside or considered unimportant as ecological convictions slipped into ideological. The singer/songwriter Ernie Constance made a public response:

Where the Snowy Mountains Rise

The old time mountain stockmen in the foothills still are found
With mobs of baldy cattle where the old snow grass abounds
With horses and with stockwhips they seem from long ago
Stockmen on hardy horses all reared there in the snow.

...

Now in the pub at Jindabyne side by side they stand
The trendy skiing tourist and the drawling old stockman
And the tourist stares in wonder as the old man tells his life
Of life in the early days where the Snowy Mountains rise.

Before they dammed the Snowy and flooded that old town
With tears in eyes he tells how old Jindabyne was drowned
Now they pump the water westward to irrigate the land
And there's people waterskiing where the old town used to stand.

Come see the Snowy Mountains, the tourist brochures say
You'll see so much you like here, you'll probably want to stay
But the changes they see coming make the stockmen grieve
One by one the stockmen all roll their swags and leave.[28]

Granville Crawford made his own defence. To the charge that the Orroral valley farmers had walked indifferent from their farms, Granville Crawford replied that neither they nor he had:

walked off owing to failure, but because of Government Resumption and in some cases a leaseback with insecure tenure. The farmers loved and respected and worked with the land and the environment ... To say that these people failed is incorrect. Fine merino wool and cattle were produced on a viable basis. Orroral valley operated by the Gregory family and Gudgenby by the Bootes family, with its magnificent Hereford cattle are two classic examples. As for people walking off; there are still families at Naas Valley that have remained for four generations ... My wife and I moved to [Sherwood Park] because of resumption and insecure land tenure, not because of any failure.[29]

By 1990 Granville Crawford had lost all three of his areas of deepest attachments—the valley lowlands, which were becoming infested with noxious weeds, the bridle tracks of the Brindabellas, and the airy plains of the high country. From Sherwood Park his mind wandered to the sites of his youth.

The feeling of terrible loneliness I suppose, a craving for familiarity ... I may be an oversensitive bloke, but memories come back every day. It might be the Coolamon Plain, up on the Snowy Mountains, it could be Murray Creek in behind Mt Bimberi, Gudgenby, or riding at Bobeyan with Ted Brayshaw when I was a little boy. Could be rabbiting over on the Murrumbidgee River.

Did his sheep really damage the mountain domains? Granville Crawford made his private answer to the critics in January 1984:

I've smelled the sweat of a wild bush horse
Felt the thrill of a reckess dash
As we've wheeled the mobs to the mountain yards
To the tune of a greenhide lash.

The mountains at first were my cradle
They later became my home
Now the bosses who live in Canberra
Tell me that in them I no longer can roam.

I've seen the mists and the fogs in the valleys
I've seen the beautiful sunset hues
As they change with the season, the mountains
To beautiful mauves and blues.

The wildflowers that bloom in the mountains
Come forth with each new spring
How they survive through the harshness of winter
Is truly a remarkable thing.

They tell me the hooves of my horses
Or the trample of cloven feet
Will kill all these beautiful flowers
But this idea sure has me beat.

For a hundred years now and better
Stockmen have used these hills
And as I look over the country
I certainly see no ills.

And now as I enter my twilight
I think and look back on the years
I'm forbidden to live in these mountains
And my only response is my tears.[30]

PASTORALISTS AND ABORIGINES

The affections of pastoralists are cross-cut by Aboriginal attachments. The farmers occupied country from which the Aborigines had been displaced: did this not weaken their own claims to belong? The echoes of religious and spiritual affiliation in the pastoralists' poetic language reminded outsiders of analogous Aboriginal attachments. If lovers of wilderness found their own attachments hard to reconcile with pastoral values, could the Orroral valley farmers legitimately share their love for the land with past and present Aboriginal owners?

Geographically different regions were generally distinguished by Aborigines as spiritually or socially different. Thus the plains to the east of the Canberra–Cooma highway, the farming valleys and the eastern escarpments of the Snowy Mountains may have been recognised as having their separate customs, extended families and Dreaming (creation) sites. Aboriginal country was differentiated, among other means, by the different languages which were spoken on it: probably at least three distinct languages or dialects were spoken in the area now enclosed by the Namadji National Park. Social connections extended far beyond the divisions of language and kin. There were ceremonial and cultural links between the areas now known as Eden, Bega, Braidwood, Tumut, the Upper Murray and Gippsland.[31]

Some of the same country inhabited and loved by Granville Crawford was inhabited and loved by the Ngunnawal people. Ngunnawal country, as far as can be ascertained, extended through the country between the Murrumbidgee and Tumut Rivers, and included the areas now known as Michelago and Canberra. Before white settlement the population density probably was not great, though greater than the whites—perhaps one person to every 70 sq. km.[32]

Probably a number of individual family groups moved within and associated themselves with Granville Crawford's country.

Aboriginal peoples lived in Granville Crawford's country from between 14 000 and 20 000 years ago, and in the northerly points a little earlier.[33] The lower ranges were in part the product of 'fire-stick farming';[34] the lower valleys such as Orroral probably were open gum and peppermint forest interspersed with grasslands.[35] The Aborigines utilised some of what has estimated to have been 200 species of insects, fifty species of birds, thirty species of snakes and lizards and twenty-six species of mammals. The manna gum, the candlebark and the red spotted gum produced edible mana. Plant food included roots and tubers, seeds and seed pods, immature cones, berries, stem and stem pith, leaf buds, mature leaves and fungi, orchids, lilies, bracken and yams. Larger animals included the tortoise, pygmy possum, grey kangaroo, broad tooth rat, echidna and wombat. The population of these declined in winter, which caused a corresponding drop in the human population, to perhaps between 5000 and 6000 individuals.[36]

The pastoralists came to the high country in spring and so did the Aborigines. In caves and crevices throughout the winter months the Bogong moths lay torpid by the millions: 'In October as soon as the snow had melted on the lower ranges small parties of blackfellows would in fine weather start for the rocks on the summit to get Boogongs.'[37] The ranges were famous for annual gatherings. Hundreds of Aborigines came for the feasting lean and hungry and returned to the lowlands in February well-fed and sleek. Dozens of observers wrote about them. The moth gatherers came from every point of the compass, including the country familiar to Granville Crawford—Uriarra, the Molonglo plains and the Brindabellas. Some of the eastern Aborigines probably followed the access route later used by the Orroral pastoralists, those regions soon to be known as Glencoe, Gudgenby, Bradley's Creek, Yaouk, The Circuits, Murray Creek and the Lone Pine Trail, past the blue haze, deep shade and grassy plains where Crace and Oldfields Huts shielded the mountain men from the wind and the rain, and on to the home of the Bogong. This land is the soul-home of Granville Crawford; it is deep Aboriginal country also.

Non-Aborigines have sometimes justified their occupation because the Aborigines had gone from the land, but in 1994 the historian Ann Jackson-Nakano identified ten contemporary Ngunnawal extended families, who mostly identify with country to the north-west of Canberra.[38] There are many hundreds of people of Ngunnawal descent mostly living in towns and cities within the old homelands, in Yass, Burrowa, Canberra and Queanbeyan, and beyond the old language areas, in Tumut, Eden and Cann River. Much less is known about the ways in which the Ngunnawal and other peoples formerly related to the land which Granville Crawford loved. A famous Ngunnawal who knew the Orroral valley well was Hongyong, who with his group of sixty to seventy people occupied the mountains beyond the Murrumbidgee and the plains of southern Canberra.[39] We know which white settlers the Ngunnawal lived near or worked with, the names of some of their wives, husbands and children, and where some were buried—but we know almost nothing of their feelings for the land. Did the Ngunnawal, at their deaths, regard their country as lost, or would it be always Aboriginal land from which they were only temporarily debarred?

Aboriginal attachments were formed by more than a physical presence. Aborigines of the Victoria River district (NT) acquired rights to country from their parents, from the woman who gave birth to them (and her sisters and brothers), and the man (and his brothers and sisters) who acquired rights to those children by marrrying that woman. These were quite specific rights of ownership and belonging, which took precedence over other rights acquired through marriage, residence, willingness to learn and demonstration of responsible management or knowledge.[40] The birth of a baby in a certain place conferred certain rights to the site, expressed, for example, as mythological knowledge of how it was created, which could only be imparted by a relative who had that same right to knowledge.

Land throughout Aboriginal Australia was created by ancestral beings. Each valley, hill, rockhole, waterfall, group of boulders and even prominent rock was likely to have been part of a complex mythology, involving its creation and its relation to the physical and non-physical world. In Granville Crawford's country, Ngunnawal and other Aborigines told creation stories, now mostly lost, to explain how their land came into being. Possibly they were analogous to this verse from an Arnhem Land song cycle, in which two spirit-beings, forming the country as they travelled, are described in the act of creating a sacred well:

We shall just plunge our *mauwulan* point in, making country,
Making a sacred well for ourselves ...
Yes indeed, Waridj, at Banbaldji we put in the point of the *mauwulan*, making ourselves a well.
When this is done, we withdraw the *rangg*, and our water comes up from the well:
Comes rising up, spray splashing! Water coming up from that *mauwulan* point.
Yes, *waridj* Miralaidj, leave it now. This well is ours! We must cover it up: it is sacred.
No one may see it, that sacred well, no one come near it!
For the water comes roaring up, spraying and swelling
From putting in the *mauwulan* point, water rises, splashing and foaming, from inside the spring.
It is our well water that roars; *wairidj*, we cover it up for us.
Yes indeed, that is good. Thus we shall go along, making the country.[41]

In the valley lowlands, the ranges and the high plains of what is now the Australian Capital Territory, there were sacred sites, stone bora (ceremonial) rings, dangerous sites, sociable places, locations where certain events had taken place, women's areas, men's areas. There was recognition of separate rights of belonging: ownership, marriage, residence and the right to speak about ceremony and law. Some parts of Granville Crawford's country may have been avoided by all Aboriginal peoples, certainly some parts would have been visited by only certain categories of people such as mature women. Those with responsibilities to maintain and care for certain sites probably carried out, or directed, increase or acknowledgment ceremonies. At Yankee Hat, just a few kilometres from Gudgenby homestead, Aborigines camped, painted the walls of the granite monolith and formed stone tools. They probably used the area as a base camp to gather moths on Mt Kelly.[42] On a peak still unnamed by the non-indigenous, but no more than 30 km as the crow flies from Gudgenby, lies a stone arrangement which determined and well-informed bushwalkers can reach only after seven hours of difficult hiking. Over 40 m long, and oriented south and north, three mysterious lines of stones on sloping granite slabs seem connected to huge rocky outcrops. Granville Crawford did not know of it; today nobody knows its origin or purpose.[43]

Aboriginal relationship to land was essentially a two-edged responsibility. It had to be cared for by its owner or owners, and the country had a duty to nourish and sustain its people. Aboriginal country knew who was there, who was trespassing on it, who had the right to be there. The anthropologist Debbie Rose summarised some of the responsibilities of Victoria River Aborigines towards their country: keeping the country clean (burning it off properly); using the country by hunting, gathering, fishing and generally letting the country know that people were there; protecting it by not allowing other people to use it without asking; protecting the country; particularly Dreaming sites, from damage; protecting the species related to that country; protecting dangerous places so that harm did not come out of that country; providing a new generation of owners to take over the responsibilities; educating the new owners in the knowledge and responsibilities towards the country and learning and performing the ceremonies which kept people and country in harmony.[44]

Are the attachments of Granville Crawford and the mountain pastoralists in some ways comparable to those of Aborigines? The land use patterns, as D.J. Mulvaney remarked, certainly were: they were contrasting but equally humanised social, economic and ideological systems.[45]

Let's review the claims which the pastoralists made to a spiritual connection to their land. Leona Lovell claimed that 'The normal mountain person feels exactly about the land as the Aboriginal people do ... we don't own the mountains, they own us ... it's just a feeling that you belong to it.'[46] A supporter of the Victorian pastoralists wrote 'The mountain people share with the Aboriginal people a deep sense of belonging. The High Country is their sacred ground. Cattlemen refer to their grazing runs as their "country", not just their "land". They identify completely with the mountains and regard themselves as being a part of the natural environment.'[47] Granville Crawford reflected: *I don't know how much the Aborigines loved this country, all I know is that I could not imagine them loving it any more.*

Granville Crawford's knowledge of Aborigines is drawn more from books by Ion Idriess and A.P. Elkin than from personal experience. In the Orroral valley Aboriginal material culture in the 1930s was no more than a curiosity. He noticed the waterworked stones on a granite ridge at Glencoe Station and wondered why the Aborigines had brought them there. He turned up grindstones while ploughing near Nass and stored them in the shed. Herbert Oldfield showed him what was known in the 1930s as 'the corroboree ground', a cleared space at the edge of a ti-tree flat near the Cotter River. Today, overgrown and forgotten, its position is known probably to no one. In all his life he met only one Aborigine, a shearer.

Granville Crawford readily concedes that Aboriginal knowledge of the country was greater than his: pastoralists lived on introduced food like rabbit and trout, for example; Aborigines spent all their lives in the bush. Nevertheless there were analogies between their care for the country and his. Like traditional Aborigines, Granville Crawford feels alienated and unhappy away from his country. He survived there not by technology or intensive labour, but by skills, experience and an intimate knowledge of the land. He followed analogous management practices like burning the grasslands for renewed growth. He could recognise a multitude of species and name certain sites of personal or public significance. Living and working in the country endowed him, he believed, with moral responsibilities to protect certain species of plants and animals. He had an intimate affinity with animals, especially horses and dogs. He sprang from the 'chosen people' of that country who knew it, spoke for it and cared for it. He was taught by older men; he felt he had a right of belonging conferred by birth, memory, long residence and family; and his sense of belonging was sanctioned by others who shared those common rights. Land was something to be loved, enjoyed, lived in and thought about, but it was also a duty, a care, a business and an employment. Pastoralists like Granville Crawford and Leona Lovell loved the mountain pastures to the depth of their beings. They came perhaps as close as any non-Aborigine to Aboriginal perceptions of and attachments to land.

Granville Crawford is no descendant of the literary current which carries alienation and rootlessness as a normal or expected part of Australian life. In philosophy he is closer to the American philosopher Emerson, who wrote that 'Spirit'—the Supreme Being—did not 'build up nature around us but puts it forth through us'.[48] Granville Crawford sees his closest connections with his country expressed through an intimate knowledge of it. The bush is to be used, though with respect. Like many Australian farmers, he sees no inconsistency between rural beauty and rural utility.[49] Though no transcendentalist, he yearns like the American naturist Thoreau 'for one of those old, meandering, dry, uninhabited roads ... where no farmer can complain you are treading down his grass, where travellers are not too often to be met'.[50] His deepest response to the Orroral valley, the ranges and the alpine pastures is close to Wordsworth's lines in 'The Prelude':

> Above, before, behind
> Around me, all was peace and solitude,
> I look'd not round, nor did the solitude
> Speak to my eye, but it was heard and felt.[51]

Granville Crawford wrote:

The freshness of the eucalypts after a thunderstorm is something that has to be tested by one's senses. It could never be described. But the feeling of great peace which descends upon one when alone in these mountains brings with it a realisation of our great creator which cannot be experienced anywhere else. I offer my thanks for being privileged in having this great experience.[52]

Whatever complex emotions were and are held by Aborigines towards the Brindabellas and the high plains, it is clear that they are not identical to the cultural expressions of love and awe felt by pastoralists. Aborigines lived on a different metaphysical plane. They recognised no intellectual barrier between the physical and non-physical: a boulder might simultaneously be a boulder and a spirit being. In this enormously intellectualised world, animals spoke their own language, spirits were objects, trees cried, the earth wept.

Who loved this lost country more? Philosophically the question is meaningless. Both Granville Crawford and the Ngunnawal loved the country to the depth of their beings, but their beings were culturally formed and framed differently.

More to the point are the political differences. The high plains national parks are out of bounds to long-term Aboriginal residents as surely as they are to the mountain pastoralists, but Aboriginal people do not see the land as lost in the same way as does Granville Crawford. It was theirs, they say, and it is still; they are merely forbidden to use it.

Albert Mullett, a Victorian Aboriginal elder whose country extends to the high plains of Victoria explained:

I don't think we would look at it in the context of being dispossessed, because to us our country is always our country, and no one will ever take that thing away from us. How it affects us is not enabling us to go and do things that relate to our culture and our traditional customs, accessing areas which we have used for thousands of years for collecting food, for cultural and spiritual purposes relating to those areas, or for recreation; because under state and federal legislation the national parks are locked up. They say they are available to the general public. Not so much dispossession but accessibility to these places with those needs. We are not allowed, under whitefeller law, to go and practise our traditional customs, access traditional food, because it's their law.

To give an example on the south coast of New South Wales there's a place called Mystery Bay, between Narooma and the mission at Wallaga Lake. We've been going there with our uncles and aunties, or taking my children, as they took their children. It's a sort of a neutral ground meeting place where for a very very long time, Koori [Aboriginal] families have been going there for food area, resource area from the water. And now the local government has implemented a policy that it's a 'primitive camp site' so it's only fine to have forty or thirty-eight campers there—So they've denied the Aboriginal communities—we have to book in now to camp there, and we're saying bullshit. What they've done is shut us out.

[In these and other areas] if we can go and make arrangements with the Ranger we can go and camp there and do the things we need to do, but we are restricted, right? And these fellers went with their uncles, spearin and camping in there, in that time we could go in there and spend weeks there whatever we wanted to spend in hunting for their learning, we can't do that now, there's a time frame on how long we can stay there.[53]

Albert Mullett's son Colin did not believe that non-Aboriginal Australians could ever be as attached to the land as the indigenous peoples are:

Not in the same context as the Aboriginal people. Australian European people have lost contact with the land over the last 2500–3000 years ... There's no comparison.

I can understand non-Aboriginal people who maybe come from a farming background and have a relationship with the land, and they see a similarity to the land they're using with where they came from, and they become very attached and very spiritualised as well, I suppose, to that country, and feel they're protecting and looking after it; but in realistic terms they're destroying it.

But world wide there's an awareness of destroying the country, and people are becoming more concerned about it ... I respect [that] in this context, yeah I do think they develop a spiritual relationship with the land.

How, if at all, can this lost country be shared? How (runs an implicit or explicit argument) could or can the pastoralists legitimately love country which is a national park, and which in any case is still Aboriginal land? At the Namadgi Visitors Centre exhibition, most information is presented about the natural phenomena of the park itself: 'Wilderness implies remoteness, great beauty, undisturbed nature—a legacy from the beginning of the earth. The Bimberi Wilderness is such a place.'[54] A film shown continuously during 1995 depicted both Aborigines and pastoralists as having once occupied the park: both historic eras, by implication, came to a natural end. The human inheritors of contemporary Namadgi were shown to be the bushwalkers and skiers.[55] Another area of the exhibition that presented information about the pastoralists was dispassionate if not cool: 'European settlement in the mountains of Namadgi represents only a brief period of history. Even so, pastoralists have left a changed landscape: the valleys of Namadgi have fewer trees than before and the remains of homes, huts and fences mark an isolated pioneering life.'[56]

However, signs of their existence in the park have been so removed that except for several sites in the Orroral valley, the pastoral past has almost ceased to exist. More information is presented about Aborigines, and while signs of their life in the park are almost invisible to untrained observers, children are invited to imagine the Aboriginal past: 'When you are out in Namadgi, keep an eye out for any good boulders that you might think might be a place that Aboriginal people camped at.'[57] No invitations are extended to imagine the life of pastoralists.

In this way the current inheritors of deep attachment to the areas enclosed by the national parks and high pastures of southern Australia have diminished the attachments of the most recent occupants. They fail to appreciate the affection of people like Granville Crawford for country which is now lost to pastoralists more permanently than it is either to environmentalists or to Aborigines. We have still to learn the nature, and think through the consequences, of multiple profound attachments to country which can be emotionally shared, at one level or another, by many generations. Australians have yet to understand how all these deep attachments can and must be accommodated in the national ethos. Margaret Johnson remarked about Windermere Station, *You were lent this land and this experience and you should be prepared to move on. You have to let it go.* In the end, perhaps everyone is only lent the country they love.

CHAPTER 4 | TWO DEAD TOWNS

People who remember Old Adaminaby recall a picturesque village on the banks of the Frying Pan Creek in the southern highlands of New South Wales. Disbelieving the news in 1957 that the town was to be inundated, a town resident wagered that he would be able to cross the creek at Christmas without getting his socks wet. By then the crossing was 20 m under water. Old Adaminaby had become one of several Australian villages submerged beneath artificially created lakes. The mining town of Yallourn, in the Latrobe valley of Victoria, is remembered more easily by former residents because its demolition was not complete until 1980. Yallourn was much bigger, better designed and more productive, the campaign to save it was better organised. It too was destroyed, by an enlargement of the open cut mine upon which its existence depended. In this chapter I consider in particular the attempts by outside authority to undermine residents' attachments to these vanishing sites.

Adaminaby was settled by nineteenth-century Kiandra goldminers after the workings cut out. The miners became squatters, trappers, shopkeepers and farm labourers, and their names were commemorated in the street names of the old town. After the Second World War, just before the advent of family motoring, the 500 inhabitants were still almost as remote from the rest of Australia as they had been half a century earlier. Seven or eight gates had to be opened and shut between Adaminaby and Dry Plains only 15 km away. Canberra, over the mountains, was four days ride away. Cooma, two days away, was not much bigger than Adaminaby. Some of the older residents, it was said, had never been to either. In the 1950s 1–3 ha farms were common in the village, on which the residents maintained geese, a hayfield or a cow. A barter economy and larger farms of 15 ha by the creek kept many families in the Eucumbene valley out of debt without regular employment. Adaminaby, for most of the year, was slow-moving. People regularly visited their neighbours, or walked up for afternoon tea in the cafe each Saturday afternoon. On the famous Saturday nights, stores would remain open till 8.30 p.m. Annual balls were staged by the Red Cross, the Country Women's Association, the Race Club, the Returned Servicemen's League, the Church of England and the Catholics, often preceded by the Juvenile Ball after which no one would eat for a week. There were six football teams, three cricket teams, concerts, fortnightly dances, a three-day show, a three-day race meeting. Adaminaby was the social, commercial and religious hub of a district which included the towns of Berridale and Jindabyne, and twenty-five rural properties which depended on the town for supplies. In good seasons the economy boomed. It was not uncommon for a station owner to fill up a 3 ton truck at shearing time.[1]

Residents of new Adaminaby recall the old town mostly with affection.

We didn't have the conveniences but we didn't look for anything better; a beautiful old town, a lot of fruit trees, pine trees, a football pitch; my grandmother had a wonderful flower garden, an orchard, a herb garden; a wonderful lifestyle; our town was a good old country town; a quiet town; a happy town; we were not in each other's pockets.[2]

Non-residents recall the winter wind howling up the south-facing slope towards the town, ink freezing in the school inkwells, being snowed in for three weeks at a time, children at Mass without socks, newspapers lining the walls of tin shacks.

Yallourn was as planned a site as Adaminaby was unplanned. Construction began in 1921 for a town which would provide an attractive living area for the State Electricity Commission workforce needed to produce brown coal and power. As for Canberra, with which Yallourn had many similarities, an architect was commissioned to design the town. A.R. La Gerche admired the English 'model city' movement, especially Welwyn Garden City in Hertfordshire. Among the stringybarks and messmates on the rolling hills between Moe and Morwell he designed a town which would enable Commission employees to live close to, yet not be dominated by, the vast open cut brown coal mine. In his plan, long straight roads were replaced by a civic centre with streets opening onto a town square. La Gerche planned an initial construction of fifty small homes of pleasing vistas, uniformity and clipped grass, within easy walking distance of work and protected from the prevailing winds. Parks and recreation areas were to occupy a quarter of the town area.[3] By the 1950s over 5000 people lived in the town. The famous cinema was the most magnificent in rural Victoria, the number 1 sports oval one of the finest in the country. The Melbourne Symphony Orchestra played in the town hall. In thirty years Yallourn became the industrial, commercial and cultural centre of the Latrobe valley.[4]

Behind each civic success resided the steady patronage of the State Electricity Commission. Familiarly known as the SEC, it provided the deciduous trees for which the town was famous, designed the English cottage style of the houses, and manufactured the distinctive yellow bricks and the red or green tiles from which the buildings were constructed. It provided soil, plants, gravel, timber edgings and manure to relieve the stiff infertile clay. It maintained the roads and services. The SEC made Yallourn, in the opinion of town planner Helen Weston, one of the brightest but briefest flowerings of the garden city movement interpreted in Australia.[5]

Remembered with less affection is the deep whine of the power station turbines and the constant rain of coal dust from the factory which turned the pulverised brown coal into briquettes. Almost everyone who lived in the town endured one or the other or both. Most vexing was the dust. Yallourn was affected by two prevailing winds, and when the wind was from the west the town was covered with a fine rain of particles. Forty tons of dust to the square mile translated into ten or twelve bags of coal dust above the ceilings of every house, which SEC workers periodically came to vacuum away. New residents were advised not to touch their washing when it was wet, but to bang it dry before they brought it inside. Most parents became adept at removing specks of coal dust from their children's eyes and to turn their eyes from their inky bare feet. Day and night, Yallourn residents heard the grind of the dredgers digging out the coal and the whistle of the train taking the coal to the factory and power plant. Every year the sound was a little louder; every year the open cut mine approached the town a little nearer. In the 1930s the SEC had discovered that Yallourn itself was built upon the brown coal seam which gave the town its purpose. Opinion differs on whether from that moment the SEC planned the town's eventual destruction.

Almost all the former residents of Yallourn, like those of Adaminaby, think of their former town with affection. In a radio talk-back show in 1994 a caller remembered a line of baskets outside the butcher, marking the place in the queue of half a dozen shoppers; another recalled the herring-bone pattern of red and green rooftops. Others called it beautiful; solid; gorgeous, built to last.[6] There were local events recalled with enthusiasm, such as the annual time trial ascent of the briquette chimney on Boxing Day.[7] Hilary Shaw recalled Yallourn was like one big family—everybody knew one another.

I can remember my dad getting his banjo out one warm summer night and all the neighbours would come out and sit on our lawn. We had community singing and they did the barn dance and waltz on the street. People cared for one another. If anyone got married the whole town would go and have a look, same for a funeral. I don't think the word 'town' suited Yallourn, a 'village' was more like it, with beautiful gardens, street trees and a band rotunda where the Yallourn band played every Sunday afternoon.[8]

There were criticisms. Some residents recalled the sharp division between wage workers and staff employees which determined children's playmates, the guests invited for afternoon tea and the composition of golf teams.[9] Other critics noted the prevailing expectation that it was the SEC's responsibility to fix everything. Leonard Stretton, the royal commissioner who inquired into the 1944 brown coal fires, blamed the SEC for failing in its obligations to free citizens:

> here indeed the townsman enjoys all that the heart of man may desire—except freedom, fresh air and independence. He lives his life on a great many days in a fine rain of abrasive coal particles and breathes with them, perforce, the nauseating stench which comes from the neighbouring paper-mill, and against which closed doors and windows offer no defence. Be he never so provident, he can not acquire a home nor an

> equity in a home. He has no authoritative voice in the management of the town, because there is no democratic local government in the town ... There is no hall where the townspeople may assemble as of right.[10]

After the Second World War the SEC went some way to answer these criticisms, but still allowed its Yallourn citizens no more than an Advisory Council. Some—probably most—of the residents were content to have their lives organised in this paternalistic way by the SEC, but at a price. The head gardener recalled wandering around Yallourn every March marking every garden with a score out of 100 points. Former coal worker Bert Taylor stated that if a worker or his family were not doing 'the right thing' (such as keeping the house and garden tidy) the SEC would 'literally take them to the outskirts of the SEC designated area and say you're not allowed [to live] in this area any more.' The SEC, he concluded, 'had a very strong authoritarian streak and not the benevolent bloody father'. A regional shire councillor described Yallourn in 1967 as 'a township of suffocating paternalism'.[11] Outsiders from Moe or Morwell either envied Yallourn's facilities, or were glad to remain outside the bureaucratic controls.

It was 1949 when Leo Crowe, who owned some of the land near Adaminaby, pulled up his horse to ask a stranger what he was staring at through his binoculars. *That's where they're going to put the dam.* Others were equally incredulous as they heard the news that the Snowy Mountains Authority intended to flood the Eucumbene valley to produce electricity and irrigation in New South Wales and Victoria. Three years later came much worse news. By diverting other rivers and resiting the dam, the Authority now proposed to create a much larger dam which would hold nine times the volume of Sydney Harbour—and inundate the town of Adaminaby. Many townsfolk maintained that such a thing could never be. Someone suggested that there was not enough water in the rivers to fill the dam; another proposed that the Authority build a wall around the town to protect it from inundation. Those in the high part of town who did not know whether their houses would be flooded were shown a marked post to indicate the future water level, then were assured that everyone would be required to leave.[12]

The debate about the destruction of Yallourn begins with whether or not the SEC or the town residents knew that sooner or later the brown coal upon which the town stood would have to be extracted. Bert Taylor thought there was no question but that the town was permanent when he arrived in 1956, but engineer Ken Murray thought that it was obvious that sooner or later the town would go.[13] The first official notification to citizens of the impending destruction of their town was as sudden as it had been in Adaminaby. The announcement was contained in the SEC newspaper *Live Wire* in October 1961, under the undramatic headline 'Playing Fields To Go First. Yallourn's 1995 Demolition'. Thereafter *Live Wire*—presumably under the SEC's editorial direction—carried on as if nothing unusual had happened. Perceptive readers realised there was more to the story from a news item a month later, which advised that newcomers to the town would have to pay rent 22 per cent higher than the current rate.[14] Anxiety was allayed by the lengthy schedule of removal. The final date was beyond the retirement time of most of the workers; there would be plenty of time to move out; and surely a whole town could not be moved simply to get at four years' coal reserves—the public buildings were so massive and so expensive; how could they possibly be demolished?

There was much less waiting time at Adaminaby. The inundation of Adaminaby, Jindabyne and Talbingo and creation of the whole stupendous network of dams, tunnels, power stations and roads was justified by the post-war national agenda of irrigation, construction and production. Public officials called on Australians to take up the challenge of the pioneers, and develop the country, or wait for invasion. A 1959 ABC radio program illustrated both the excitement of the construction and the disregard for the human cost of development typical of the 1950s. The broadcast depicted the southern rivers 'calling out to man'; the nation was 'still far from mastery of its own country'. Leeton farmers downstream were reported to be 'jubilant'. Thousands of people were promised employment, there would be four new towns and dozens of construction camps. A construction worker proclaimed the rush of men and machines to be 'the greatest turn-out I've ever seen in my life'. The human cost was presented as negligible. A teacher was heard to say to his Jindabyne pupils:

> Hands up all those whose houses will be flooded. [Most hands, presumably went up. The teacher continued.] What should the government do about houses? [Pupils answered] We should get new ones.

Listeners were reassured that plans were in hand for a new township. New Jindabyne would be 'as modern as tomorrow'. Dr McPhillips of the district hospital service related that she had visualised the lake and thought of the benefits to Australia. 'I know the benefits of power and water. In no other country have I seen civilised women working so hard without domestic or technical help. Perhaps that is the answer to make life easier for us all.'

In the euphoria of national post-war construction, few thoughts were given to the sites soon to vanish. The only reservations allowed in the program were offered by Father Coles of Jindabyne—'Although it's a national project, naturally we feel very sad about losing our place'—and by the teacher's wife who, to a background of sad music, lamented, 'At the back of my mind there were two things. The greater thought is that this is happening for the greater good of our children, the other is that feeling of loss.'[15]

Some months after the reporter's visit, a tremor of emotion is said to have transfixed Jindabyne residents watching army sappers dynamite their bridge, but the Authority's propaganda film on the destruction of the town allowed only the trite sentiments of the film's jolly title tune, 'So Farewell This Town of Mine'. So died Jindabyne. Did the residents grieve? While Adaminaby's younger residents seem to have accepted the offered benefits without much regret, the older residents who opposed the destruction of Adaminaby seem to have been made almost speechless by the gigantic scale of the Snowy Mountains project and the highly successful public relations effort directed by its Chief Executive, Commissioner Sir William Hudson. The few residents who protested they could continue to live comfortably without water, power or sewerage were unreported by an uninterested press. The Snowy Mountains Authority officials alone heard, but did not publicise, their alarm.[16]

Compared to the other lost places discussed in this book, the attachments of Adaminaby residents seem most muted. Why? Perhaps, such was the massive intensity of the Authority's program and the strength of its employees' belief in their national mission, the Adaminaby residents persuaded themselves that they did not care, or if they did, that they should not grieve. The national and local sentiments in favour of development were probably greater in the Snowy Mountains in the 1950s than at any other place and time in Australia's history. Adaminaby residents were overwhelmed by the publicity, the size of the operation and the language directed at them by the Snowy Mountains Authority which intentionally or unintentionally belittled the attachments of the villagers and rendered them, many years later, hopeless and bitter.

The Authority made much of the two new towns which it planned would replace Jindabyne and Adaminaby. For the first time residents would enjoy mains electricity, running water and sewerage in their modern planned towns. Adaminaby houses capable of being removed (in effect, the wooden ones) would be moved free by the Authority to a new town site to be chosen by the residents themselves, while public buildings like churches and banks would also be removed or rebuilt. The public relations exercise implied that those who stood in the way of national progress were selfish or unpatriotic. Commissioner Hudson reasoned that 'the loss of this historic township and the pastures of the Adaminaby basin' was 'more than justified by the enormous potential of the water'.[17] Adaminaby people, like all Australians, were enjoined to believe that modern machinery and skilled people could provide water to quench Australia's fiery heart.[18]

Besides the public relations, the might of the Snowy scheme was visually conveyed by the 18 000 visitors to the scheme annually. It was in the dozens of roadside explanations, the excited press reports, the lookouts over dam and power station sites, the sealed roads, the guided tours, the hundreds of workers who roared and caroused in the Snowy Mountains pubs throughout the mid 1950s, the heavy machinery gathering behind every hill, the new roads appearing from and apparently leading nowhere.

The potential for distracting or crushing local attachments was reinforced in the language of authority. In films and leaflets the Authority used phrases like 'it was necessary' for the dam builders to commence, as if the bulldozers obeyed some natural phenomenon like gravity. It referred to the 'loss'—not destruction—of the old town, the 'changeover' and the 'smooth transfer' to the new, as the old one 'submerged'—not died. A pleasing inevitability was implied in the assertion that the Snowy River was 'destined to play its part' in the nation's life, that Australia had reached the 'end of one era and the beginning of another'. The 'transfer' was frequently presented in the passive voice even when the Authority was the agent: 'the site was chosen, procedures were explained, difficulties were overcome'. An Authority film entitled *Operation Adaminaby* cast the removal of the buildings as a military-style heroic rescue: helpless Adaminaby was not to be 'left to the rising waters'. The sole concession to emotion was the image of a 'Road Closed' sign beside a road leading into the water, with the comment 'They will pass this way no more'.[19] Even at the dam face where 'a saga of toil unfolded', workers who began the destructive work were neither named nor even identified as workers, but subsumed into the generic 'Australia'. The project was 'necessary', it was 'more than justified'. The Authority had drawn up its proposal: an orderly and irreversible process was implied in 'the decision taken, planning began'. As the waters rose the racecourse was not destroyed but 'disappeared'. At the new site the town was 're-established' and the people 'allocated' new houses. A sedentary state, long recognised by governments as a desirable condition for citizenry, was appreciated in residents who 'settled down' and 'established gardens'. A section in a booklet which was to have been published by the Authority was entitled *Settling In*. It was difficult to challenge an Authority when sitting down and 'settled'. The author's first draft attempt at reasoning was replaced by a second draft assertion of scale and power: the justification that the land to be irrigated would 'more than compensate for the pastures now disappearing beneath the waters' was amended to 'will dwarf the area of the pastures now disappearing'.[20]

There was not much of the national agenda that could be offered in favour of the destruction of Yallourn. The SEC's arguments were purely economic. The town was ageing, the coal was under the town, it had always been intended that the coal should be exploited: failure to do so would cause unemployment and blackouts. Engineer Ken Murray believed that the town had served its purpose of siting workers near production: as modern transport enabled workers to live further out, the town was expendable. The most economically rationalist position was that 'in principle nothing should be allowed to interfere with the production of coal mining with its allied operations of electricity and briquette production to the utmost economic limit'.[21]

Community anxiety, lulled by the faraway date of 1994, returned when the SEC announced that, contrary to its earlier prediction, it would 'begin to remove' the town almost immediately. A 'Save Yallourn' committee was formed in November 1968, which for perhaps the first time brought staff and wage workers together in a common purpose. Prue McGoldrick, a long-time resident, wrote that 'No town, especially one as beautiful as Yallourn, can be obliterated from the face of the earth without some form of protest, and Yallourn was no exception'.[22] At a protest meeting in June 1970 the Yallourn residents sang together, to the tune of 'Yankee Doodle Dandy':

> We are the people of Yallourn
> Our hearts are overflowing
> It makes us sad the news is bad
> To see our town is going
> Let's gather round the old town square
> Our voices raised in protest
> We'll sing its praises loud and clear
> To save it from the dozers.
> So let the old town carry on
> Around the square we'll rally
> And sing its praises loud and strong
> To save our home town.[23]

The Save Yallourn Committee persuaded the Victorian government to ask its Parliamentary Works Committee to investigate the planned destruction. For four months in 1970 evidence was presented by the SEC and its opponents. The townsfolk argued for a new pricing policy, that four years' coal did not warrant such hurtful destruction, and that the coal quality was so poor that oil had to be added to make it burn. A.D. Spaull, an academic and former resident, argued for Yallourn's status as a garden city and its successful industrial relations. Yallourn symbolised 'the attempts to eradicate the evils of the century of laissez faire industrialism and replace it with a new mode of social organisation; to destroy it was an act of collective vandalism'.[24] The inquiry concluded that the destruction of the town was justified.[25]

In 1956 the Adaminaby townsfolk were offered the choice of where New Adaminaby was to be sited. Perhaps dominated by the business proprietors, they did not choose the obvious position just above the old town, but an open, windy site 8 km away (convenient for businesses as it was on the main road to Tumut). Nita Stewart, who still lives in the new town, recalls being ferried about on a freezing July day, when everyone was so cold and miserable that they didn't really care which site was chosen. The events to most people were simply bewildering. Nita Stewart now asks herself, *We had rights, why didn't we use them?*[26] At an angry meeting in 1956 Hudson had challenged the townsfolk to name any organisation that had gone further than the Authority in giving the people a fair deal. Rex Yen claimed that the town representatives had approved the site, but not the people. Several voices supported him. 'It wasn't. You're right. The councillors were too ... weak.'[27]

Authority officials began to assess the potential of each dwelling in Adaminaby for removal to the new town. Wooden houses were to be moved for free. Owners of brick houses, which could not be moved, were to be compensated on a pro rata basis calculated on an estimated 100 year life of the building. Furniture storage, temporary accommodation and removal costs would also be paid by the Authority. The residents were allowed a choice of three home sites in the new town. A woman was filmed apparently enjoying a cup of tea with the Commissioner on the verandah of her home, perched atop a moving semi-trailer. Nita Stewart recalled of the event:

It was done as a publicity seeker by the Snowy Mountains Authority to show just how kind the Commissioner was to the people of Adaminaby. The people were all asked to line up for the photo to be taken to prove how popular the Authority was. [The owner] was asked to cook scones on the trip out to prove these houses were everything the Snowy had said they were.[28]

Memories converge on the last few months of the old town. First to go was Bill Crowe's farm by Frying Pan Creek, then Mrs Carter's 40 acre farm, then the sportsground and racecourse, then Ken Mackay's place, followed by J.S. Mackay's, then Geoff Yen's, then the Ecclestones', Chalkers' and Dave Mackay's.[29] In the wet spring of 1956 roads turned to mud as residents stared at the house next door grinding past their front door, then at the empty house site. Soon a second space opened on the other side. The diminishing village began to resemble the open valley slope it had been a century earlier. One by one the shops closed, some to reappear in the new town, some to vanish altogether. Older people who were teenagers at the time recall the excitement of removal, how they persuaded their parents to throw away the heavy water fountains, iron bedsteads, sulkies and furniture for which there would be no need in the new all-electric, sewered and watered town. Nita Stewart remembers she and her sisters throwing down the well—because the tip would soon be under water—the mundane possessions of a lifetime: pictures from the walls, caketins, ornaments, hats and shoes. Her mother, Nita Stewart now realises, watched in mute protest. A farmer abandoned—and still remembers that he abandoned—a huge camp oven, stoves, glass lamps, beds, a sulky. Others speak of a Buick and a truck unaccountably abandoned beneath the lake. Whether they were scavenged or covered by the water, these lost possessions are now a matter of intense regret.[30] Some residents didn't care: the sooner they moved out of the old town, they said, the better.[31]

The anguish of those who cared can be gauged by the events of the last few months of 1957. At the Eucumbene dam the water is rising at a foot (30 cm) a day. The last balls, church services, dances and picnic races are over. Owners of houses not already moved have been given the date to be ready. The post office closes in September; 'New Adaminaby' becomes 'Adaminaby' and 'Old Adaminaby' officially ceases to exist. Electricity, which has been connected by the Authority for its own purposes, is disconnected in October. The *Cooma Monaro Express* comments 'Although several families still remain in Old Adaminaby, the main street is deserted, shells of former buildings are waiting patiently for the rising waters to claim them, and the loose rubble from many foundations tells the story of the big move.'[32] In November the 101st (and last) house is removed. An Authority photographer records the event in black and white: the foreground depicts house and semi-trailer; the background inadvertently reveals the dying town—empty footpaths, collapsing verandahs, naked interiors, piles of debris, exposed ceilings.

By December the farms and orchards along the river are gone and the water is creeping towards the town site itself. All that remains are outhouses, foundations, chimneys, immovable stone cottages, roads and trees. Lanes, dusty for a century, are tinged with green from disuse. Fruit trees have their best but briefest season. People dig roses out of their gardens for their new dwellings. They calculate the height of the water by watching its progress up the pine trees at the back of the recreation ground. All at once the pines die. The fruit trees too die before they are covered. Among the sharpest memories are those of frequently returning to the water's edge to stare at that which soon will be nothing. They stand in fascination, dismay, anguish or horror before the scummy water inching towards what remains of their homes. It's lapping at the back fence, creeping up the garden, swirling about the outhouse, eddying round the lintel, muddying the fireplace, splashing about the chimney. Oblivion. *We were surprised at the levels which went under first. Sometimes it'd come up so quick, and you thought it was weeks away.*

As the waters approached the village, film-maker Robert Raymond directed *A Town to be Drowned*. Unlike the Authority film-makers, Raymond explored the images of destruction: a ruined building, a dismantled street, a fuel stove alone on its mounting, a broken windmill, a fence and gate half submerged, water lapping at the edge of a stone cottage, garden and sulky, trees three-quarters inundated, a bath in place amid broken walls, shoes, bottles, kitchenware, a staircase leading nowhere, a boy crying. Over and over Raymond returned to the visual theme of the rising water, oddly agitated, black and sinister. Interspersed were images of brutal bulldozers and high-pressure hoses. Adaminaby was *Waiting for Death. Death by drowning.*[33]

Yallourn residents too watched in alarm as, from the early 1970s, one by one the wooden houses were jacked up, sometimes split in half, and towed away. At first the removals were violently opposed:

> The 'Save Yallourn' committee had only a few hours notice, that the house in Fairfield Avenue was to be removed. The Committee rang up their people and asked them to come along early in the morning and try parking their cars in the street. It meant that the semi-trailers couldn't get in to take the house away, a silent protest, so to speak. Then the police declared the street a no-parking zone and woke up the local residents who had parked their cars there inadvertently.[34]

The Save Yallourn Committee appealed to the Central Gippsland Trades and Labor Council (TLC) which threatened a strike if so much as a sink was removed.[35] Removals continued. The SEC's answer was that the town population had already dropped by nearly 1000, that few residents would be required to move out and that the houses were thirty years old and needed replacing: 'We should plan for the gradual attrition of the town.'[36]

In January 1974 the Victorian premier declared that 'the town must go'. In April the federal Minister for Minerals and Energy declared himself appalled and promised to help. In May, candidates for the federal election were asked to state their position on the town's destruction. The Gippsland TLC called it 'a terrible act of vandalism' and an 'act of sheer madness to destroy a town of 1,000 homes', and placed a black ban on the demolition.[37] But in June the SEC, seemingly as inexorable as the waters of Lake Eucumbene, advertised for the 'demolition'—not 'removal'—of eight brick buildings. On 30 September the carefully named Yallourn Resettlement Committee invited residents to afternoon tea to meet officials and to fill out a form stating their 'intentions after removal'.[38] Monash University students squatted in the empty houses. In July 1975 another TLC meeting was advertised.

This May be your Last Chance
Take No Notice of Rumors, Half-Truths and Assumptions
There is Still a Chance of Saving This Beautiful Town

The SEC replied that 'Yallourn must go', that it was 'too late to turn back current plans'.[39]

Public feeling reached its highest point on 29 July 1975. A huge barbecue was held in the town square; stories of it are still recounted with enthusiasm and nostalgia. Older residents returned for the occasion, some after many years' absence. Inside the library was a photographic display, outside were concerts and a dance. Bands played, people reminisced, made speeches and carried posters reading 'Old King Coal Has No Soul'. Two songs were composed. The first was 'Welcome Back to Yallourn':

Welcome back to Yallourn
May a smile or a word bring your memory delight
Welcome back to Yallourn
Special town, special job giving power and light

When you see childhood faces from childhood places,
The warmth of a welcome brings you closer once more
If it was home or the place where you were born
Sincerely welcome back to Yallourn.[40]

In a much-quoted speech, Joe Dell, the president of the Save Yallourn Committee, took the rostrum and began:

> When I was a little boy I took great delight in stirring up ants' nests and then controlling the movements of the little creatures using a stick, sand walls and water and this way I could control where they went very well, in fact I could make thousands of them go around in a circle wondering what happened to them. Over the last few years I've begun to realise just how those directed ants felt.[41]

This famous gathering was the biggest protest, and the last. Thereafter the townsfolk lost heart rapidly or were attracted elsewhere. Long-term householders were paid a disturbance compensation, provided they went quickly. Residents who had bought their houses to transport elsewhere objected to union electricians refusing to disconnect the power. The *Express* reported that 'the time for saving Yallourn had passed. Its population is drifting away; and every municipality in the Latrobe Valley is, at this moment, preparing to pick out the eyes of Yallourn's magnificent community facilities.' The reporter was referring to the recommendation of an independent arbitrator that the town's treasured facilities such as the pool, the theatre and sport fields should be rebuilt by the SEC in surrounding towns.

At this prospect, non-Yallourn residents had little reason to prevent Yallourn's immediate demise. Only 200 people attended the mass rally at which, on 5 September 1975, the black bans were lifted.[42]

The officials of the SEC almost matched the Snowy Mountains Authority's language of bureaucratic inevitability. It described the necessity to 'win coal from beneath the town', which therefore must 'go', be 'abandoned, vacated' or 'removed' by 'gradual attrition'. The people were to be 'rehoused' in new sites to be 'allocated'. The authorities implied the scientific necessity of 'destruction' because there was 'no alternative but to go through the township' by the 'extension of the open cut through the town'. Yallourn itself was devalued as less than a living town in its characterisation as merely a 'town area' or a 'community'. 'No useful purpose', it seemed, would be served in 'retaining' this 'commission asset', this 'community asset' or this 'burden'; it was an inconvenience to 'maintain and service'. The town would 'transfer'; the people would 'leave their homes' and 'set up elsewhere'. There was a suggestion of the unsavoury in SEC verbs like 'eliminated' and 'razed'. While there was little heroic about 'planned obsolescence' or 'a town built to go', a greater manifest destiny was implied in the SEC audio production entitled *Sounds of Power*. Like the Snowy Mountains Authority's 'end of an era', Yallourn was depicted as a 'star' which had 'reached its zenith; brown coal [read Australia] has decreed its demise and Yallourn must disappear'. The more mealy-mouthed pronouncements referred to 'the current plan, the disappearance of Yallourn by a far-sighted realistic attempt to plan for its [the SEC's] future needs', and, most remarkably, the 'planned retirement' of the town. The management-dominated Yallourn Town Council, in its final meeting, thanked the SEC for allowing the 'demise' of the town to be as 'orderly and dignified as [could] reasonably be expected'.

The attachments of Yallourn citizens were anything but crushed beneath the weight of SEC's linguistic devaluation. The Save Yallourn Committee described the plans as 'SEC strangulation and unwarranted vandalism', while the TLC asserted that 'the destruction of the town [would be] the greatest act of vandalism in Australia ever witnessed'.

The townsfolk noted the SEC's intention to 'do away with' the town. They spoke of houses being 'taken' and 'coming down around us'. The SEC was 'emptying people out' by 'destroying' the town and 'digging it' or 'clearing it away'. They watched 'the last house go down'. While some maintained that the SEC was entitled to 'take it away' or 'evacuate' the people, there was a sense of acute loss in the image that the coal cutters would 'move in and cut it [the town] up'. To such people it was 'a very very traumatic experience'.

Unlike Adaminaby, Yallourn was very often personified. The analogy to the death of a person was obvious to everyone except SEC officials. Yallourn had been 'condemned from the day it was built'; it had been 'born to die' from the first, and it was 'a sorry thing' to see a town 'die' or 'in its death throes'. Some residents likened themselves to trees, and when the SEC chose to 'rip something out like that you rip out people's roots'. There was a hint of Christian resurrection in a town 'quietly disappearing' to have a 'rebirth elsewhere'.

Those whose business was with words evoked the destruction of the town most powerfully. Ted Hopkins (the poet 'Slab') introduced to his book of Yallourn stories, 'outrage, died, destroyed, ruined, degenerative forces', and 'crushed'. Other writers returned to the town site, appalled by the 'devastation', trees 'savaged' and 'the ground levelled'. Mick Thomas, a former resident and member of the band Weddings Parties Anything, wrote 'Industrial Town' which personified the town as a genderless 'old grey friend'. He invoked the SEC to 'dig you up, tear you down', and, in the meditative beginning to the song, to 'lay the town over, dig it all in'. The former residents are portrayed as looking at the 'crevice', the 'grave', where once the town lived. Now they 'cry for a ghost town'.

Memories hang heavily on the last months of the town. Bernadette McLaughlin and her parents had moved from Yallourn in 1967, but found in the early 1970s that the town continued to *draw us back*. On Sunday afternoons they would drive in from Traralgon to see friends in the town, play in the park, admire the deciduous trees or have a swim. They always drove past their former house. By 1971 it was apparent that its turn for removal was approaching. The family visited the current owners to have a last look round. It was *very sad. With it went all my memories. I'd like to take my children back and say 'this is where I had my childhood' but I can't, it's really sad*. After the last family visit, her father, still working at the Yallourn factory, reported fresh developments to the family: *Now the roof's gone.* When the dwelling was removed he returned to dig up some roses and camellias. The family bought timber from the demolished high school and a load of the distinctive Yallourn bricks from the town centre. Bernadette McLaughlin is not sure exactly which buildings they came from, but it doesn't matter. What is important to her is that the bricks came from Yallourn. *The buildings were part of your life.* It was about this time that the family stopped their Sunday drives to the site. *There was nothing left to go to.*[43]

Another youngster was Alan Lucas. His parents had left when he was a child but, like everyone, it seemed his future lay with the SEC. He was employed in the engineering department which, like the Snowy Mountains Headquarters, continued to use the town site after its community's eviction. He watched *the heart of it go. The shops, they were the first.* One or two individuals, like the milk bar proprietor, fought on to the end. One day Alan Lucas passed the half-wrecked cinema, once the pride of the Latrobe valley. It was where he had had his first date, where he saw *Midnight Cowboy* and *The Godfather*. He clambered over the wreckage and souvenired a copper number from the back of one of the seats. *It was a solid memory or connection.* He found it disturbing driving round the deserted streets, bereft of houses and signposts, listening to his friends identify trees or house sites as the scenes of romantic adventures. Watching the town die like this, he thought, was *like watching a friend die of cancer*. Was it better that it should happened fast or slowly?[44]

David Andrew returned to the site in 1979 to find electricity trainees taking lessons in the library and sheep grazing on the school playground. He found his family home vandalised and a barbed wire fence running through the middle of what had been his mother's treasured garden. The front gate and brick fence remained; he had himself photographed standing at the site.[45]

For a time the economy of New Adaminaby prospered. The Snowy Mountains Authority offered plenty of work. A population explosion of trout feeding on drowned grubs as the waters rose attracted hundreds of fisherfolk. By 1975, just as the Save Yallourn movement reached its peak, Adaminaby's economy began to slide. Jindabyne had the ski traffic, the timber mill was closed, Cooma monopolised the fishing festival, the trout population had declined, and the receding waters in dry years left the lake's edge muddy and unattractive. What had been a twenty-five minute drive to Jindabyne now took an hour. For recreation, religious observance and commerce, the families on the other side of the lake now turned towards Cooma. Few had foreseen and none had cared that the lake would destroy the social networks of a century. Cars and the sealed road made it easy to bypass the town for cheaper supplies in Cooma. Skiers from Mount Selwyn dashed through to make Sydney by midnight. Rates were payable to the Authority on all services instead of the former single unimproved land tax. The Authority granted building licences at Angler's Reach, a popular campsite 15 km away, but Adaminaby residents were ineligible to move without the waived interest payments on their relocated houses in the new town immediately falling due. Electricity was expensive. The running water frequently had to be boiled to be drinkable.

The Authority, allocating itself the role of an SEC-style benevolent dictator, is said to have regulated the height of fences, the position of water tanks and the permissible number of chooks. The old timber houses did not cope well with the strain of moving; skilled workmanship was rare when they were built. Doors sprang open unexpectedly, windows refused to close, floors creaked, a teapot might be shaken off a dresser by a footstep, chimneys smoked, doors misaligned, roofs leaked. Mrs Elizabeth Ecclestone was given a fowl-house door for the toilet and when she complained was asked 'aren't you ever satisfied?' Hardly anyone remained in their original houses. A retired couple might be given an older, smaller one with an inferior stove while the family which moved into the larger one was indignant that it was now in debt for the supposed improvements. The re-configuration of buildings long identified in certain relationships was worrying. A resident might now look across the road at what had formerly been the nuns' residence at the other end of the village; next door might be a house which used to be three blocks away.[46]

So the great weight of the Authority continued to crush all those affected by the enormous engineering achievement. Most of those displaced were, and seem to be still, gripped by the conviction that since their sacrifice was in the national interest, their attachments did not matter. A moving lament about the drowning of Talbingo town under the Blowering Dam begins with the familar concession 'Sure it's a mighty job of work, give credit where it's due.'[47] Almost all the twenty survivors of Old Adaminaby who were interviewed for this book acknowledged the importance of the project. The national agenda, expressed particularly by the language of bureaucratic inevitability and rectitude, presented them with a narrow range of emotional options to assuage the total destruction of their living areas. A sense of grief, or tragedy, or poetry was flattened by the official cliche 'the end of one era and the beginning of a new'. In its thoughtless assumption that it was better to start life afresh than to allow time to grieve, the Authority ineptly tried to encourage community morale in garden competitions whose chief effect, in local memory, was jealousy and bitterness.

Not all the Authority's activities, though, were accidents. Community memory, as the historian Connerton noted, is sometimes the struggle of memory against state power and forced forgetting[48] and the authorities did their best to obliterate traces of the old town.

The Snowy Shire Council tried to change the name of the new town to Chifley. The Snowy Mountains Authority marked Coolawye on maps where Old Adaminaby should have been, and discouraged signs and memorials where the now deserted upper part of Denison Street slides gently into the silent lake. Housing blocks became available above the old town site only to those who lived more than 30 km away, effectively preventing those in the new town from applying.[49] Not a single person, to the knowledge of one resident of the old town, received an invitation to take part in the Authority's fortieth anniversary celebrations in 1989. Why was it, Nita Stewart asked in 1995, that the Snowy River Shire made not a mention of Adaminaby in its 'Mountains Master Plan?'[50] She concluded a letter of protest to the *Cooma-Monaro Express*, 'Yes it was at ADAMINABY that it all began, but perhaps to the "powers" that be the "oldies" are still back in the Old Town, under the water perhaps.'

The sentiments of the Authority seem contained in this lament by a poet for the drowned Welsh town of Tryweryn.

> Nothing's gone that matters—a dozen farms,
> A hollow of no great beauty, scabby sheep,
> A gloomy Bethel and a field where sleep
> A few dead peasants. There are finer charms
> Observed in rising water, as its arms
> Circle and meet above the walls; in cheap
> Power and growing profits. Who could reap
> Harvests as rich as this in ploughmen's palms?
> All's for the best—rehoused, these natives, too,
> Should bless us for sanitation and good health.
> Later, from English cities, see the view
> Misty with *hiraeth*—and their new-built wealth.
>
> All of our wealth's in men—and their life's blood
> Drawn from the land this water drowns in mud.[51]

In the early 1990s, the older residents of New Adaminaby felt that they had been duped even in the first meetings called to discuss how—not if—they would move from the old town to the new.[52] The mood of the present town is one of disappointed resignation. *I could show you the spot tomorrow*, or *mixed feelings now, a bit of sadness*, or *I don't know how to explain, it's just that we were happy there and satisfied to stay put*. They are particularly unhappy with the slow decline of the new town: *we had two- or three-day race meetings, now you have a job to get one*, or *you can get burnt here striking matches looking for people* or *we don't feel at home in [these] houses, they're not as homely*. They feel that their particular burdens were that they were the Authority's guinea pigs, that their economy has stagnated, that the population is ebbing, and that their beautiful site above the old town is occupied by strangers.[53] Outsiders who watch the slow death of the new town pinpoint the malaise: *to a certain extent they're grieving; they're in deep mourning; a very severe shock and they never recovered from it*. They point as evidence to one of the service stations which used to close early, to the dusty old general store, complete with the original cedar counter from the shop in the old town, with hardly anything for sale. The residents of Old Adaminaby have not recovered from the loss of their town because they were never allowed to mourn it. There are few picnics beside the old town, no pilgrimages, no heritage walks, no wreaths floating, no museum, no map, not a solitary sign beside the scattered half-dozen houses and the glittering lake. This is not the slow peaceful death of other country towns.

Nor were there plays, films or community theatre for the Adaminaby residents, only dreadful reminders of a past assumed to be forever lost. In 1983, during the worst drought since the dam was filled, the water level dropped so far that those who had lived in the top half of the town saw the foundations of their dwellings reappear *like a corpse from a grave*, as one resident put it. Flowers bloomed in the same beds in the same gardens where they had grown twenty-five years before. People knelt in prayer in the ruins of the old church, had picnics in their gardens, were photographed standing on the muddy steps. The process of uncovering has now occurred four times, so that people describe the reappearance of the town in reverse order to the inundation:

The streets are clearly defined, the gardens are soft, you can drive anywhere in the old streets when the lake goes down. The kerbing outside the RSL club, that normally comes out, and the back of the Roman Catholic Church was one, and the twelve steps up to it appear, then Yen's store, and Granny Kennedy's tree comes out, a fine old pine tree.[54]

In 1983 many residents found the experience *very dreary and depressing ... I didn't like going back to a ghost town and see the road suddenly disappear and go to nowhere.* Nita Stewart recalled:

I went back because one of the places I really liked was where my grandmother lived, and she had this big acreage and she had a tennis court that had a lot of happy memories for me ... Anyway, I told my children about this tennis court, and then when the day came and the water was down enough for us to go and find it, it was ... so sad, and then as I stood on the tennis court I could see grandma's cow-bail where as a child I'd had to run around that paddock to put the calf in. And even the gate was there and the fence, and you know, you could actually, just like going back into the past and seeing something: you know, if there's been a calf in there I could have even chased it in and said 'Now that's what we used to do.' And her orchard. That also, the trees were dead, naturally, but we used to go over as children and help our grandfather pick the apples ... and even the foundations of where her shed was, and the house and the pine trees, and yes I found that very very sad, and I know it was really terrible for those who actually owned the house to go and see it.

It was dry. As a matter of fact, after being years under the water, the sand was still on the tennis court, and the fences, that was the amazing thing, the fences were still there, even paling fences, they were still standing there. And chimneys. And that's a ghastly thing to see the chimneys. A lot of them, when the lake was going down, they'd hate to hear that because they'd know what they were going to see.[55]

It was the outsiders who best articulated the reflections of the residents of the old town. Marge Mackay knew Adaminaby but lived on a farm which was not inundated: her diminished sense of loss perhaps enabled her to sense the emotions which the residents of the town could not put into words. At the conclusion of interviews, I often read the former residents part of her poem 'The Old Town'. One verse runs

> The stations, farms, the sheds, the barns
> Are all awash in the deep waters
> While ghosts of men swim through the glen
> Drowned faces haunt their sons and daughters.[56]

In the stillness a very common response was *Yes, that's it exactly.*

Yallourn was different. Some sporting bodies in the nearby towns of Moe, Churchill, Morwell and Newborough retained 'Yallourn' in the names of their relocated clubs. At Yallourn community feeling was better co-ordinated and more articulate. A plainly expressed economic motive was easier to confront than the claims of a national interest. The residents have reinforced each other in their denunciation of the SEC and in the articulation of their attachments. Dozens of people contributed to the book co-ordinated by the Save Yallourn Committee, *To Yallourn With Love*. Ted Hopkins's *Yallourn Stories* is underwritten by the observation that

> Slab loved Yallourn because more than any other place
> Yallourn was an intelligent arrangement of form in space.[57]

A schoolgirl, Urzula Horbach, wrote in greater anguish:

> And as she [Yallourn] is destined to be torn asunder and
> The riches gouged from the dark recess of her soul,
> She is abandoned, sacrificed by her children
> For time has clouded their vision of her humble role
> And she is mercilessly forsaken
> By man whose insatiable greed has led
> Him to condemn her with his threat of power.
> And through the deserted streets where life has fled
> Her spirit awaits the agony of inevitable death.[58]

There was fun, too, in the attempted (and failed) levitation of 125 million tonnes of coal by an individual whose previous claim to psychic powers was making a teapot fall onto its side. There was a mock-heroic pageant intended to be performed before Princess Margaret:

> Actors dressed as SEC officials begin to bury Princess Yallourn in her coffin in view of Princess Margaret on the lookout tower. After much fuss and shouting, the people of the valley are shocked by the cruelty of the SEC. They turn against the SEC officials, driving them away from the intended burial. Princess Yallourn then awakes from a deep sleep.[59]

Connerton's argument about 'forced forgetting' carries less force at Yallourn. The SEC was unable to crush or devalue memories of the town. The 1975 'Back to Yallourn' was as much a celebration of community values as a funeral. The film *Born to Die* commissioned by the SEC gave inhabitants—unlike the Adaminaby residents—opportunities to speak about their feelings. 'It's not born to die it's born to live and it's living now; Isn't this a town worth keeping; How much is a town worth? The present is all I have; Just dreadful I think, just dreadful; a very big break to make.' The last shot in the film was a close-up of the dredger blades chopping at the grass, approaching the town ever nearer.[60]

Another opportunity to re-articulate community attachments to Yallourn was a play written in 1988. A modest plan to dramatise some industrial disputes in the town's history became in 1989 a full-scale amateur re-enactment of the town's birth and death. Pat Cranney, a producer and author of community theatre, wrote with the help of the former residents a powerful script of song and dialogue. The story followed Everywoman (Jean), who comes to Yallourn as a young married woman. She does not like it at first, but her husband refuses to leave. The keystone of the play is Jean's refusal, twenty years later, to move from the town at her husband's request. She replies:

> But you had a dream—it was Yallourn. And it's happened, just as you said it would. And we are part of it. A dream come true—that's more than a lot of people get in their lifetime. This is my house now, Peter. Nobody's taking it from me. I belong here. And I'm staying.[61]

The Yallourn Story, which ran a full season of sixteen packed performances, traverses the formation of the Save Yallourn Committee, the Public Works Inquiry, the famous Celebration of 1975 (in which, in most performances, Joe Dell played himself), and the final moments of the town. Jean sweeps the floor for the last time as she resists attempts by her family to leave for her new home. A chorus of ghosts, accompanied by drum beats, chants:

> A dying house in a dying town
> A dying house in a dying town
> And memories are ghosts that won't lie down.

These are the last words of the play:

JEAN Just go and let me be, will you. I'll be ready by the time you get back.
JOHN Come on.
But she can't even sit down. The house is empty.
[Kathy and John exit]
JEAN Empty ... Empty she says. Ha Ha. Fifty years of memories. Empty!

The curtain falls on Jean still sweeping the floor.[62]

Yallourn people bought their houses and re-erected them elsewhere, or hunted for them in nearby towns and photographed them. They dug out plants, they souvenired bits of their houses and the public buildings, they bought building supplies, and laid their paths with Yallourn bricks. Alan Curtis lives in Canberra because, in part, Canberra's planned streets and autumn leaves remind him of Yallourn. Sociologists who interviewed many former residents concluded that they had not suffered much pain or grief.[63] Though this judgment was perhaps a little superficial, Yallourn residents probably were less traumatised than Adaminaby people because they were allowed emotional space to speak and write about their feelings. They have continued to do so. They were more numerous and better educated than the people of the Snowy towns, and the mass rallies, protests, committees, interviews and the official inquiry gave them opportunities to gain confidence in their own sense of belonging. In the 1970s, green bans and struggles to preserve several well-known buildings had already alerted Australians to the symbolic and emotional value of the physical past. The SEC was easier than the Snowy Mountains Authority to portray as a capitalist, greedy, uncaring juggernaut.

My enduring image of Adaminaby is a still, cloudless day in midsummer, a slight chill in the air, a deserted potholed road slipping into the silent lake. Roads at the site of Yallourn also end suddenly at the edge of the open cut, but the central image is what everyone calls 'the big black hole'. Viewers at the lookout site stare into the vast black coal face, the dredgers far below, dust high in the air, and reflect that somewhere, in that empty space, was the town of Yallourn. Prue McGoldrick, like almost all the former residents, returned once on the journey to nothing.

Not long after we had removed our house and some things in the garden, we returned to the town. It was still possible to gain entry and cruise nostalgically along the beautiful tree-lined avenues. When we reached the top of Reservoir Road where our house had stood with its back to bushland, I was shocked to find the trees gone and the ground levelled. Because it was some distance from the developing big hole [open cut] I assumed it would be spared such treatment. It was unbearably sad and I cried. I could not look again at such devastation, even if it were possible.[64]

CHAPTER 5 HOME: THE HEART OF THE MATTER

Homes, like other places, are mentally constructed. What we identify as 'home' is not only a different location from everyone else's, it occupies a different space. Home can be an area as big as half of Sydney:

> Dad knew the city tracks. Not just the steps and pathways round the Cross, for example, but he had a mental picture like a map. The short-cuts all the way from the coast to Parramatta which makes me think of Sydney as like a middle-eastern city, multi-layered and only readily knowable by people with that ancient knowledge.[1]

Home can be the inner city:

> But still the centre of gravity is the inner city, and oddly enough it is here, in my corner house, with traffic on two sides of me, that I've begun to learn how to be still, and to accept that changes can come in small and undramatic ways.[2]

Home can be a suburb:

> It's me. Footscray is me. I know I'm happier here than I've been for years ... I felt as if I've come home ... I liked it very much, I do, and I won't be leaving here.[3]

Home can be a house:

> Well, it may sound a bit corny, but to put it this way, when Helen and I went down to our place in Cherwell fifty-odd years ago, I thought that

> was the loveliest place that anybody could ever have. It was a nice brick home that I had and I think home is everything; you've got to put a lot into it and you get a heck of a lot out of it.[4]

Home can be a room in a house:

> [When someone was in the kitchen] it kept the family in contact throughout the day. When they're home, everybody knows where to find other people, or at least to find Mum and Dad or whoever's doing the cooking—there's usually somebody in the kitchen.[5]

Home can be a single plant in a garden:

[My attachment] is to houses. The big weatherboard house in Campsie which we sold and was knocked down to put up units. It's not even the house, I think it's actually the back yard. And the flowers in the back yard that I'm attached to. So it's bluebells and snowdrops.[6]

As well as the space it occupies, people conceptualise their home as the functions it performs. To some, home is a comfortingly bounded enclosed space, defining an 'other' who is outside. Others, more socially attuned to their neighbourhood and friends, see 'home' not as a place but an area, formed out of a particular set of social relations which happen to intersect at the particular location known as 'home'. 'Home' can be a focus of memory, a building, a way of mentally enclosing people of great importance, a reference point for widening circles of significant people and places and a means of protecting valued objects.

'Home', as T.S. Eliot remarked, 'is where one starts from'. The loss of a loved place sharpens perceptions of what is most valuable in the shaped and fashioned space. The affection for a home, in western cultures, is the point where griefs for lost countries, towns, properties, gardens and suburbs seem to meet. Home is the ultimate focus of all lost places.

On 16 February 1983 the temperature in southern Victoria rose to 43°C. In the ranges west of Melbourne people were particularly nervous. A bushfire a fortnight earlier had come perilously close to the towns of

Macedon and Mount Macedon; water was desperately short. On this blustery day a second fire from the north seemed to have bypassed the town. Nevertheless many families had packed their cars with precious belongings and were awaiting the signal from the fire station to evacuate to the nearest shelter. For families in the top end of the town the designated shelter was the brick Macedon Family Hotel.

A few hundred metres from the hotel lived Peter and Sue Boekel. They were schoolteachers at Gisborne, but rented their timber dwelling at Macedon from the Education Department. It was an older place, two bedrooms and a sleepout. They had been in the town a year and were beginning, as they said, to put their roots down.[7] Macedon seemed a sleepy town. Peter Boekel joined the amateur dramatic club, but the Boekels were not pub-goers and, apart from some of the parents of children they taught in school, did not mix much with the townsfolk. On 16 February Peter Boekel was returning from a school excursion, exhausted by fractious children and an overheated bus engine. Sue waited for him at home in the thickening smoke. It was obvious from the confused media reports that a national calamity was occurring—but where, and where next? From the fragmentary news it seemed likely that the fire would bypass their town, but Gisborne, downwind from the prevailing gale, was threatened. Should they evacuate? The Country Fire Authority had advised them to wait for the signalling siren. A fire truck arrived to evacuate one or two families: that was puzzling as well as worrying. By 7 p.m. it was apparent that the wind had shifted from the north to the south-west. The 3 km front had widened to 15 km and was racing towards the town at 60 knots. The windshift had probably saved Gisborne but Macedon, in the path of the freshening winds, stood directly in the altered firepath. The Boekels' home would be among the first to be consumed if the wind did not abate. Still there was no word from the firefighters at the station, who, had the townsfolk known it, were occupied elsewhere or were cut off by the flames. Smoke and cinders increased. At 7.30 p.m. Peter and Sue Boekel, alarmed by the ever-thickening smoke and the ominous roaring over the hill, decided to make for the designated safe zone. As they hurried to the Macedon Family Hotel the tops of the pine trees upwind from their home caught fire. They parked their car, piled with their most precious personal items, opposite the fire station, then scrambled into the hotel.

Across the road, in the Macedon post office, postmaster Paul Gray sat at the Country Fire Authority two-way radio. There was no longer much that he could do. The fire station was empty and the major firefighting equipment was cut off at Bacchus Marsh: his post office was now the headquarters of the Macedon fire service. The town was defended only by a tanker and a couple of trailer units parked outside the hotel. People urgently radioed to ask what to do. Paul Gray replied that they should judge their own circumstances: if they had the equipment and personnel to save their homes they should do so, if not they should leave at once.[8]

Paul's wife Eleanor was calming the large number of people taking shelter in their house. Two of her three children, their tiny treasures already in the car, were ready to run across the road to the hotel, but the third, Yvonne, was keeping company with an expectant mother in a nearby house. The power failed and Paul switched to the battery, amid the undefined but menacing roar from the direction of Gisborne. The wind was fierce and the sky was boiling red as Eleanor gathered up her elderly charges and escorted them across the road to the hotel. She returned for the kids, prayed that Yvonne was safe, and stood by the hotel window, watching her darkening home and pondering the wisdom of returning to fetch the wedding albums. Where was Paul? Where was Yvonne?

Paul Gray continued to receive desperate messages. At about 7.50 p.m. the room was abruptly lit by a weird orange glow. The oak tree overhanging the house was on fire. *This is Macedon Station closing down. I'm going over to the pub.* Paul Gray ran across the road and hammered at the locked doors of the hotel.

Eleanor Gray saw her husband arrive. Worried sick at Yvonne's disappearance, she stared out the window as the wind howled and the trees bent almost horizontally. The fire station was ablaze. Unlike Peter and Sue Boekel, whose house was on the far side of a rise, they could see the post office directly opposite. Before her eyes *that beautiful old building*, ninety years old, their own home, began to burn as embers from the blazing fire station swirled around its timbers. *The first corner hurt terribly. It didn't actually draw me but I had to look at it. I think I had to believe it. I forced myself to watch it.* The oil tank exploded with a roar that could be heard above the howling of the firestorm.

Three hundred people were sheltering in the hotel, but even at the height of the conflagration people continued to batter at the door.

Ashes, cinders and burning sticks swept in with each arrival. So thick was the smoke that those not involved in urgent business lay on the floor. People drawn to the windows saw their cars, all laden with precious belongings, burst into flame. Peter and Sue Boekel remember flames blown parallel to the road, burning asphalt, smoking cinders from the ceiling ventilators swirling round the room, footsteps thundering on the roof as desperate firefighters fought to save the hotel. By 10.30 p.m., watchers saw most of Macedon in ruins and the fire racing towards the town of Mount Macedon.

News came that Yvonne Gray had sheltered safely with her neighbour. Shortly afterwards Yvonne arrived, white-faced and silent. The first crisis was past. A cool change arrived with a few spots of rain and the night grew very cold. The danger of fallen power lines made leaving the hotel dangerous. In Sue Boekel's recollection, *the mood was subdued as people realised they'd been devastated and we started to contemplate what we'd lost*. The community emerged from shelter to find a forest of chimneys and twisted tin. Six people had died, and more than 400 homes in the two towns had been destroyed.

At first light Eleanor and Paul Gray drove to Gisborne and soon after returned up the melted road to the wreckage of their home. The intensity of the fire had destroyed almost everything. A single saucer survived of a tea set 100 years old. Two intact plates remained inside the dishwasher which had crashed from the burning kitchen floor into the basement. A glass percolator with a melted handle. A glass saucepan still standing on the stove, which shattered when picked up. There was a macabre humour in recognising objects twisted to unfamiliar shapes. Golf clubs were bent double, the chandelier lay in the ashes, at first unrecognisable without its fittings. The garden swing survived, and half the boat, but the caravan was reduced to four mysterious silvery wraiths. The post office safe seemed intact but when opened by a locksmith next day the money and documents were found to be only ashes.

Peter and Sue Boekel were also returning to nothing. As they made their way through spot fires, fallen trees and solitary chimneys, *it looked like Hiroshima*. They met a neighbour whose house was still standing, then a firefighter with streaming blood-red eyes. All that was left of their home was the front fence. The water in the plastic fire buckets had evaporated to a third of the volume, and the buckets melted to the waterline. The lawnmower was reduced to an extraordinary aluminium puddle. That object must be the sewing machine. That must be the glass preserving jars, melted together. *Oh look, there are those pennies which were on top of the oil heater.* Some miniature ceramic clogs were blackened but unharmed. After half an hour Peter and Sue Boekel returned to the hotel to see if anything could be salvaged from their burnt-out car. Later they extracted some half-melted jewellery from the boot. They started to walk past the crackling power lines to seek help in Gisborne. A shire engineer drove past, ignoring their appeals to stop. A few minutes later they met him again, now confronted by a fallen tree. They helped him move it, then shared a ride to Gisborne. That night in Sue's parents' house Peter Boekel recalled *I hit low point*. Sue Boekel's worst moment was telling her mother the news on the phone. *It wasn't very nice. We cracked hardy but we were in shock.* [Later] *My mother told me—we weren't aware of it—she could tell by our eyes.* The shock of their lost home was mingled with the trauma of the event, the perilous escape, and relief at the low death toll. Sue Boekel was still in shock when she went to the Red Cross the following day for replacement clothes, and found that she couldn't be bothered looking and neither could she remember Peter's size.

Their home, never very secure in their affections, was emotionally lost to them:

Sue Boekel: *[There seemed] all the more reason not to take a photo. I wasn't interested, I just wanted to move out. There's no point in staying here, the facilities are poor, it's going to take ages to get back in, people are going to be sorrowful and I didn't want to be part of that sort of feeling ... Why hang around?*

Peter Boekel: *We were out of it, and our neighbours were living in caravans for a year or so while their house was being built, and then the rains came and it was muddy.*

Sue Boekel: *For me it was self-defence. You just start again somewhere else. That's all happened and you don't want to go back and live in that deadness, so why put yourself through it.*

Peter Boekel: *Yeah, it was a full stop. Bang. It was finished, that's for sure.*

Sue Boekel: *Those people who say, 'Oh we're going to stay, we're not going to let it lick us'—that didn't interest me at all. [For example] people who owned* *their own homes. We did go back and have a look around, but I don't get a pleasant feeling going back. There wasn't anything to see back at Macedon. There was just people building and it was all blackened, and the attraction originally was the bush, which wasn't there any more. There's nothing there to draw us back now.*

Peter Boekel: *It was good before the fire and bad after.*

Paul and Eleanor Gray were among those who chose to remain in the town. Friends lent them a caravan. Paul put the matter prosaically: their return to a functioning post office was an urgent business necessity. Eleanor had stronger feelings. Though the ruin was *a horrible charred mess there was this longing to be back on our block.* Within two weeks the block was clear:

I wish it had been longer. I remember the [fire brigade] boys came in, they had this great big tractor, and they were going to pull [the chimney] down, and they hadn't even asked us or anything, and I was so angry and I went out there like a mad woman screaming 'leave it alone!' That was all we had, just a chimney. It's ridiculous now, but that was how I felt. That they'd come onto our block. It seemed we weren't anybody, we just seemed a piece of paper being processed. As if someone died and the body's taken away and it doesn't belong to you.

She felt herself numb and unable to make decisions. When taken to buy new kitchenware she thought of the century-old crockery lost in the fire and had to be persuaded not to choose the first articles her hand fell upon. She was in shock, she thinks, *probably for months and months. You can operate day to day, but it was very hard to get out of bed and face the day.* The Grays erected a screen to protect themselves from sightseers; even so, at Easter hundreds of cars streamed bumper to bumper past the ruin. Eleanor Gray felt humiliation, anger and victimisation as she heard the cries. *Oh there's nothing, no ruins for us to look at.* Andrew Riemer had found that native-born Australians could not share the intensity of his grief for Budapest Jews. Eleanor Gray found that this charred ruin, her own place of deep personal engagement, was also a messy site for others to exclaim over, photograph and forget. Her chimney, a blackened obstacle to the tractor crew, was also the site of collective family memories, a symbol of the past, a reminder of the terrifying trauma and a protector of their precious and fragmentary possessions.

Some residents, like the Boekels, left the town and did not return, but others shared what Eleanor Gray called the *return to the nest syndrome.* She believes that the meetings to discuss the future architecture of the town helped the process of collective grieving. But grieving for lost homes in the end is personal and lonely: the most precious possessions lost were the family photos, the record of twenty-five years of weddings and babies. *We're scarred emotionally, but I'd rather be alive than have photos.* Like the Boekels, the Grays feel less drawn to material possessions. *Who cares? Warmth and company survive.*

THE RIP THAT WILL NOT STOP FLOWING

Marie Pitt, a poet of the 1930s, found that warmth and company did not survive the destruction of her family's dwelling. She revisited her childhood home in Victoria to find that neither wind, weather nor weeds had erased the pain inflicted by family members on each other. Here was no nostalgic crumbling homestead engulfed by nurturing nature:

> No roof was there, nor sign where leapt
> A hearthstone's friendly flame

No romantic ruin this, only the site of raw pain and sorrow barely forgotten, barely concealed:

> As well that o'er a green hill side
> God's own good sun spills free—
> Old orchard trees more ghosts might hide
> Than I dared see.[9]

Dorothy Hewett imagined herself revisiting her empty childhood home in Western Australia. In a vision she saw herself, her brothers and her sisters playing cards long ago:

> I will see the children
> circling the oval table
> with the dark blue inkstains,
> playing Ricketty Kate

Hewett would stand among them, listen, point, give advice: no one would hear her. And those same children would grow up to experience the horrors of mid-century existence—war, possession, cancer, the death of children, madness:

> one will be shot
> out of his Kitty Hawk
> over the Owen Stanleys.
> one will give up the farm,
> and all his possessions,
> to the Jehovah Witnesses,
> one will lose his wife to cancer
> one will lose her child
> one will be unhappily married,
> my father will be buried
> in the graveyard
> on the edge of town
> my mother will lose her wits
> but none of them know it ...

That place: the environment, the family home and the landscape, alone were unchanging:

> nothing will stay in its place,
> except the sea, the little house,
> and the black crags showing
> (above the line of the rip
> that will not stop flowing).[10]

A ruined home, especially one of personal association, speaks to almost everyone of change, decay and mutability. To Judith Wright the environment did not suggest the relentless continuity of existence, but degradation and corruption, a continuity barely worth maintaining, barely maintained. She revisited the site of the home of her great-great-grandfather:

> And the trees and the creatures, all of them are gone.
> But the sad river, the silted river,

under its dark banks the river flows on,
the wind still blows and the river still flows.
And the great broken tree, the dying pepperina,
clutches in its hands the fragment of a song.[11]

A few lost homes revive sweeter reflections. Nan McDonald returned with her sister to her family home, transmuted not ruined. Their parents were dead, but McDonald dedicated to her sister a poem which celebrated 'the lovely ghost of the living and the dead', a love free from pain, regrets or corruption. The walls enclosed loving actions and emotions which endured and could be perceived by those who had once experienced them:

The tide of love flows back,
Though she is long gone since who was moon to them;
Still shines in this suburban box of brick
The tranquil glow that once, in the next room,
Was life's first light to me, in her arms held,
Before my newborn eyes could lift to meet
That square of sky where now the whitening gold
Of the chill south-west is edged about with night.[12]

That's a second element in the remembrance of dead homes. People who have lived in the one place for many years often feel that the building knows when they are away, that it has absorbed the emotions expressed by themselves or by several generations. People leave or die, but their vibrations remain. Geoffrey Dutton wrote of his home Anlaby:

More than memories move in this house,
For in me like nerves, from nursery to deathbed
Are its rooms, and running like veins
Are the curves of its corridors conducting my blood.
Love, leave-taking, losses and hatreds
Are its population always, so if no person were there
Yet in the deep distance, the darkness of trees,
Secretly a light would be struck for my return.[13]

Memories are ghosts that won't lie down. The worst of all experiences of returning to a lost home, for some, is to find it transmuted and occupied by others who share no past history or affection, who do not feel or do not want to feel the presence of interior ghosts and memories. Marjorie Pizer returned to her childhood home to find the garden ruined and, worse than destruction, a house occupied by strangers, professional people who had no love or knowledge of its history of human engagement:

> All of this place was part of my formation
> All of this place became the background of my growing.
> The garden where I played is all torn out,
> The fruit trees gone, the grass all driven to dust

This was no home: the laundry was a laboratory, bedrooms were offices and waiting rooms:

> Ghosts of my childhood walk around with me
> While strangers work in every room I knew.[14]

Lost homes are the focus of nostalgic longing and desires. Philip Hodgins wrote:

> It's strange to drive for hours alone
> then turn the engine off and stare
> at places round the billabong.
>
> They look the same but they belong
> to other people now. That's where
> they're like a lot of things, they're gone.[15]

Edward Relph distinguished 'landscape' (part of any immediate encounter with the world) from 'place', which he thought constructed in our memories and affections, 'time-deepened and memory-qualified', a *here* from which to discover the world, a *there* to which we can return'.[16] How important is memory in valuing a lost home? How are lost homes remembered? What is the relation of treasured things to the treasured whole?

Anne Boyd, a book editor and former Dominican nun, shared in neither the collective grief nor celebration of Macedon. She lived alone in her house in Ferntree Gully, Melbourne. While holidaying in America in 1993, she received news that her house had been destroyed by fire. By the time she returned her friends had removed the china, glass and other valuables that seemed salvageable. Abroad, she had imagined that she would not wish to visit the site; she allowed her holiday to run its full course to enable her mind to adjust to the shock of a homeless return. Anne Boyd arrived at Ferntree Gully numb rather than upset. *Once I got back there I knew I needed to see it, and my instinct was to keep going back to it.*

Now began a process, only partially conscious, of separating once-loved objects into those which were to be preserved and those which were to be thrown away. Burnt pages from her large book collection were blowing all over the garden and up the street. They would have to go. Anne Boyd stepped over the step where the back door had stood: there, unnoticed by friends concentrating on objects of value, were her walking shoes, just as she had left them. They remained in the mud for the demolisher to remove. She threw away other objects which she turned up or which her friends had found: a scrap of her Dominican cloak, spoiled jewellery, pages of books which she had edited. *The burnt fragments lying, burnt and wet, fragments of my best clothes, fragments of things that had been important to me, lying in the mess.* Twenty years of letters, wet and burnt down one side, were eventually recovered, but it was only after she had been through the last box that she knew that her diary and guest book, kept for twenty years, had gone for good. So much had vanished of the things gathered over the years, which had been part of the meaning of her home: a mosaic of neighbours, former occupants, objects touched by associations, the house, the furniture, the view, the actions, the prayers, the meals, the talking.

Like the Grays, Anne Boyd found the public exposure painful. Papers and sheet music were still blowing about a month after the fire. Everyone stared as they went past. Vandals threw things around at night and wrote words about the fire on the walls of the smoke-blackened garden shed.

Many Yallourn residents re-erected their homes in other towns but lost their original home site. Anne Boyd considered herself luckier to have lost the home but saved the site. Although, like the Macedon victims, she found making quick decisions difficult, there was no question in her mind about whether she should rebuild on the same site. A process of renewal began. She cut roses from the old garden for her table in the rented flat. She returned almost every night, at first to look for objects uncovered by the demolishers, later *I just wanted to be there*. She would have lived in a van on the site if the insurance company had allowed it. The ritual of returning seemed to her to match the ritual of the Catholic liturgy; it seemed appropriate and answered a psychological need:

It was a month before the place was demolished, so that I came back frequently and walked through [the ruins] time and again, and it was like a long funeral. I felt I needed to keep coming and looking at it ... You'd walk up the drive and here would be this, this ruin. Then they took the roof off, and gradually they got it down to nothing. I felt I had to go through this process of being there and seeing it gradually go down and down until only the floor was there and the actual foundation.[17]

She began to recover from depression at the moment the builders began constructing her new house. It was to be built to her own design. In the wreckage she had recovered a blue, terracotta and white ceramic plate which was bought in Assisi and which had hung in her home in London. She cleaned it, hung it in the restored bathroom and planned the new tile design around it. *There's a pattern, a continuity, a symmetry about that.* A brass candlestick, now stripped of its lacquer, was put on the replaced piano. Anne Boyd began a new photo album which included a triptych of the living room photographed from the same place 'before', 'the day after', and 'now'. These she described as icons, *images of something important*. So were the photos unexpectedly recovered of celebrations in the old house. The views up One-Tree Hill and of the evening sky were the same as before. The carport was rebuilt exactly where the old garage had stood. The stumps and planks of the old house still lying about she kept for the slow combustion heater. A piece of ironwork wrought by a previous owner was recovered and re-erected. *I felt proud of the continuity. This was the first house in this bit of Ferntree Gully, my parents died in this area, they are buried in the cemetery here, so there is a long continuity.* There was a personal continuity too. Anne Boyd hung the washing on the restrung, almost identical, clothesline. *I found that very satisfying.* Jungians speculate whether the home is the mirror of the self, depicting how we see ourselves and how we would like to be seen.[18] Anne Boyd's house is one expression of herself, but the site of her home, not the home itself, is her place of last significance.

When Alice Rawe thinks of her first lost home she thinks first of the house, then the events, then the people. She was born in Claremont, Perth in 1935 and stayed in the same house until a year before she was married. Though she left it forty years ago, she can in her imagination still see it clearly: her bedroom ceiling, the lounge, the open velvet curtains in the hall, the heavy green cover on the dining room table, the reddish verandah, the lino on the kitchen floor. *I can remember it so well.* Within this physical configuration occur the activities and events of the home: listening to radio serials like *Dad and Dave, Martin's Corner* and the Lux Radio Theatre, sitting on the verandah steps waiting for the results of her Junior Certificate, doing jobs in the kitchen. The people take identity from the activities: her grandmother and uncle in the lounge, her mother in the kitchen, Alice herself playing 'murder' with other children. *When I first thought of this it was people, but I'm finding I can see the places too, without the people, you can separate them.*[19]

Alice Rawe left school, went to business college, found herself working too hard, went to eastern Australia for a holiday, fell in love, and married John Rawe. Until 1995 she remained in Sydney.

In Sydney the special place in the lives of Alice and John Rawe was Mosman Bay. The harbourside suburb had the same middle-class, old-established values as Claremont:

I've developed rather a strange attitude to where we live in Sydney. At first I lived in the YWCA at Mosman Wharf for a year before we married, and then we lived in Balmoral for eight years, down near Mosman, then we bought a block of land and built a home in Forestville, for four years we lived there, and then we sold that and bought back into Mosman, in Shadforth Street.

Perth, and Shadforth Street, that really felt like home to me. This home had really good vibes. [I felt] the warmth and love as soon as we went in, an instant attraction. I don't know whether you believe in things like that, but the people we bought the house from, the woman didn't really want to sell it at all, but her husband wanted to. So for five months people would come and look at the house, and she would do everything in her power to make it look unattractive. Then we came to look at it, and she felt she would want us to have it. So they waited for three months till we sold our place in Forestville. It just felt like home. Maybe because it was a familiar sort of place to the place I had in Perth. It was a three bedroom, quite ordinary sort of house, same sort of lino and furnishings.

Already bonded to the suburb through the memories of their courting—the trees, the older houses, the hills and the quiet harbour—Alice Rawe now began to sink her roots into the house itself. As in Claremont, her attachments were in the house, the events and the people within it.

The house is at the core of the district. It's very hard to separate them. The more I think about this, the more I realise that important events happening in a place make that place more memorable. I can't get away from that. In my Perth house, both my grandmother and my uncle, they both died in that house. It's all of those things that add up to it being a home, something special. I see the [Shadforth Street] house in my mind, and then I remember the things that have happened in that house. They don't all have to be important things, but some of the things that really affected me greatly, happened when I was there. I don't just mean the tragedies, I mean the good things too. I can remember getting a short story published and getting the cheque. I was there when it happened. If I look back on things that have happened, I tend to see myself in that house, where they happened. That's weird, isn't it? A phone call, I can remember that phone call when my mother died; I can remember getting a telegram from my mother saying that my father was dying. I had cancer a few years back, and I can still recall lying on the bed at the lowest point of the illness and hearing the birds singing outside. It's as clear as anything, and it's ten years ago. That happened to me there. I've had very strong feelings when I've been in that house. All these events that have happened there I remember that that was part of it all, the fact of where it was, in the memory. Now looking back I can see how very important Shadforth Street was.

Shadforth Street is the second of Alice Rawe's lost homes. It passed out of her possession not by inundation, bushfire or open cut mine, but by the mundane circumstances of daily life. By the 1980s some large trees up the back needed expensive attention, John was approaching retirement, the children were leaving home. The Rawes calculated that their superannuation should be supplemented by an injection of capital. Without agonising about the decision, they sold the house and bought a three-bedroom unit 3 km away in Cremorne. Alice Rawe did not realise what a significant step she was taking, and does not recollect much about leaving. *I don't think in those days I was aware or as in tune with my feelings as much as I am now.*

Reflecting upon the decision to sell, Alice Rawe believes now that if she and John knew then what they know now about the intimate meaning of her home, they would have remained. She continued to shop in her old suburb of Mosman as, though the vendors were different, the shops were the same. The children regret their first lost home in Shadforth Street as their mother regrets her first lost home in Perth:

The children talk about the present place as 'the unit', the Shadforth Street house as 'home' or 'the house'. All know what they mean. Here [in the Cremorne home] nothing's really happened. This place is neutral, it's convenient, it's comfortable, it's quite a nice place to come back to. I could probably live here for the rest of my life, but I have a feeling that I won't be.

I think I'd probably head for Mosman. The old barn [in the Roberts painting] is still there. In a perfect world I wouldn't hesitate, that's where I'd like to be.

Would you buy the house back again if you could?

They've put a lid on it, a second storey. No, I wouldn't, but it wouldn't be too far away from there. A better view, overlooking the bay would be lovely. But I wouldn't want to go back there. It did have problems. It felt good, and we made it good, and we were very happy there, but I don't think I would go back.

If she heard that the house was to be demolished, Alice Rawe:

wouldn't be terribly sad, we're not there any more and not likely to be, the memories would still be intact. Even though I can't bear to go down and look at it. It's not as hard as it used to be, but it looks so different now, it's got this second storey on it. [If it looked the same as when we left] I'd probably feel worse.

Though our interview had been arranged to discuss Perth as a lost city, little by little our conversation had been drawn to the lost home in Shadforth Street. I suggested that this home had been as much, if not more, of a lost place as her first lost home in Claremont. *Yes, I'm realising that too.*

A striking phenomenon in these accounts was that the men related more strongly to the district in which their home was situated than to the family dwelling. Granville Crawford, John Rawe and Paul Gray each identified with a wider area than their partners. The centre of John Rawe's world was not the home in Shadforth Street or the unit in Cremorne but the suburb of Mosman and beyond to a large section of Sydney. While Alice's world beyond Mosman was the northern harbourside suburbs—*Sydney wrapped around the harbour*—John Rawe felt comfortable *from Palm Beach to Hornsby and the eastern suburbs.* Although Mosman was the centre of familiarity and security, he felt that he could live anywhere within that larger area of knowledge:

The houses are not as important as the district or the area. I was really happy in Shadforth Street, but I can go back and look at the house, and don't have the longing that I want to live in that particular house, it's just that it's a nice street, but it's more the area in my case. Not the abode but the surroundings.

The reasons are not mysterious. Margaret Johnson and Miles Franklin sank their roots in the paddocks and bush because they were familiar to them. John Rawe's employment as an insurance surveyor kept him travelling throughout the areas where he thus feels most familiar. The cost was a lesser bonding with the family home. John Rawe was often away working when critical phone calls were received, cheques arrived or children learned to walk. His life circle was wider and more diluted.

Paul Gray was much less affected by the loss of his Macedon home than his wife Eleanor. The family lived and rebuilt on the same post office block, he believes, for no more than economic necessity. He has had no bad dreams or memories of the terrible events, and does not like the reconstructed town of Macedon as much as the old. He supported Eleanor's plan to declare an Ash Wednesday Park because it was a *dead paddock doing nothing*. The plaque erected in the park commemorates the determination and courage of the Macedon community, but gives no details of the events or the lost places.[20] Paul Gray's pragmatic feelings are expressed in his belief that life must continue. *There was a fire here once, so what?*

Men and women, then, are likely to grieve for the lost places which were most familiar to them. Women mourn lost houses more frequently than men; men more often grieve for their lost workplaces. The Angliss meatworks, in Melbourne, was to critics the site of low pay, exploitation and industrial pollution. It was hated by some male workers, but loved by others. When the factory was closed in 1976 and demolished soon afterwards, Albert Hart recalled:

I spent my whole life there, I still talk about Angliss's as though it was my other home ... I loved it and if they could reopen that wonderful old plant I would be the first in the queue of many thousands who would gladly start it all over again.[21]

George Linnard worked at Angliss for forty-five years:

I'd go back to Angliss's tomorrow if they opened up again, for sure. It was my second home, loved it. It gets pretty sad when you see it now.[22]

George Kimberley, who worked in the briquette factory at Yallourn, emotionally described the demolition of the famous chimney:

Tortured from within by the heat of fire and from without by the changing elements, it uttered no sound of distress, neither did it falter. To the ferocity of these onslaughts it merely swayed, only to recover and stand straight and tall, to again defy its tormentors.

The torture from within ceased.
Men no longer toiled beneath it.
The surrounds slowly crumbled.
Its service to man was ended.

Its work well done, its task complete, it succumbed to the hands of its creators. Watched by officials, a multitude of spectators and some who cared, it fell to earth with the same great dignity with which it stood for many years. And with equally great humility returned to the dust from which it was made.

Saddened were the hearts of many men.

No case can be developed on this evidence which characterises men as caring less for lost places than women. Men and women are likely to love and mourn *different kinds* of places. The Angliss meatworks and the briquette factory seem to represent the same emotions which others feel about their houses, determined by the time, engagement and energy expended at those work sites. The anguish felt by George Kimberley as he watched the briquette chimney tumble was shared by two women who watched an equivalent, or worse, tragedy—the spectacle of their own homes being destroyed.

TWO DYING HOMES

Pat Jackson lived at Cribb Island, a village on Moreton Bay in Queensland, which was demolished in 1979 to make way for the extensions to Brisbane Airport.[23] She and her husband had connections with 'Cribby' since they were quite small:

There was what they called the Tin Hut there, and every fortnight or so this family would come and put on a show. They'd open up the doors and have a piano, drums and violin, we'd all sit on the grass outside.

In the 1970s Pat Jackson's town had been a community of 1000 people. It had half a dozen grocery stores, a post office, a police station, a Tuesday dentist's clinic, a Wednesday doctor's clinic, a picture theatre and several churches. In Pat and John Jackson's memory, Cribb Island was a kind of paradise. *Beautiful clear water. You could grow anything. No racial prejudice that I can remember, none at all.* The children rode bikes, played softball, swam together, built sandcastles on the beach and blew up letterboxes on cracker night. Fishing, worming and shell-grit mining offered ready employment.

The Jacksons bought a block of land on which John Jackson designed and erected an L-shaped fibro house. He enclosed the patio as a sewing area. Outside was a bush house and barbecue. A friend told Pat Jackson *I love coming round to your house, Pat, it always smells of cooking.*

John would go out fishing and bring back three buckets full of whiting, we'd freeze them and that'd keep us going for months. Flathead, bream, it was a wonderful place.

Intimations of disaster came to Pat and John Jackson in October 1973. For many years rumours had circulated that Brisbane Airport might be extended to encompass the village. On 26 October the Department of Lands wrote to the Jacksons:

> ... as the registered owner of land in an area of land which the government has decided should be acquired for Brisbane's future airport development ... It is hoped to now negotiate with you a mutually satisfactory price for your property.[24]

Pat Jackson reacted:

It's not going to happen, they can't possibly shift us. I kind of thought it wasn't going to ever happen. I was very emotional, I think. I howled and cried and carried on. I was a teacher aide at the school and saw the decline from good students to kids that were itinerant.

The Jacksons ignored the letter. It was followed by another, warning them of an imminent visit by a government valuer. Failure to reach agreement would result in a recommendation that their 'interest in the land be acquired by the government by compulsory purchase'. Pat Jackson did not attend the public protest meetings. *I was a bit like an ostrich with its head in the sand. I just thought to myself, 'It's not going to happen, they can't possibly shift us.'*

As at Adaminaby and Yallourn, community life declined as long-time Cribb Island residents sold up and were replaced with renters holding neither memory of nor stake in the settlement. There were robberies and a rape, unheard of in the island's previous history. The Jacksons began looking for a new house in November 1974; two weeks later they found one 5 km away in Banyo. They concluded a deal with the resumption authorities by which they were to leave their carpets, furnishings and fittings in the house for short-term tenants. *We had searched for just the right light fitting, and the drapes had taken us ages to find just the exact colour ... It made things a bit ... heartbreaking.* John Jackson borrowed a ute and took to the new house the family possessions that they were allowed to remove. Pat Jackson *cried every night for six months when I got here, every night.*

Every week she had to drive into Cribb Island to shop for an elderly friend. She hated the trip more as the gardens grew wild and the houses more derelict. Her own house was occupied first by a single mother whom she spoke to, *but I couldn't bear to go in.* Within two years the house which they would always think of as 'their' home was progressively wrecked by a series of tenants. *John and I went down there one day and all the sliding doors and everything were all smashed.* One morning in 1979, after she dropped off the groceries:

> I came around and went past my house. And I saw it being
> pulled down.
> And I just stopped the car and I burst into tears.
> Just seeing them pulling it to pieces.
> Something that you'd strived, and John worked very hard to
> get for us.
> Just to be pulled, just to be pulled
> to pieces.
>
> Just terrible.

Fifteen years after leaving Cribb Island, Pat Jackson wrote a poem entitled 'Cribby'. The second last verse runs:

> So now we are scattered around,
> A few here can be found,
> A few have settled down
> Some are under that earthly mound.[25]

Libby Plumley spent the first twenty-one years of her life at the Weetangerra post office on the outskirts of Canberra, where her mother was the postmistress.[26] Until the early 1960s, when Libby was about twelve, the Limestone Plains seemed to go on for ever, as if Canberra, on the other side of the Black Mountain ridge, had a life independent of the district which surrounded it. Their home on its 15 ha block was the centre of the district for 7 or 8 km in every direction. Libby Plumley's nearest neighbours were over a kilometre away. She rode her bike or horse to her friends' farms, and caught a bus to Ainslie school. Her mother spent from 9 a.m. till 9 p.m. within hearing of the telephone exchange in the kitchen. Her father was a ranger. On the property he had a cow, a milking shed, a score of dogs, a dam, horses, chooks and stables, but most of the day he was away on his horse. Weetangerra was so quiet that on Friday mornings, when he drove into Civic, the nearest town, the dogs would start barking when they heard Dad returning in the old Dodge, labouring and winding around Black Mountain, half an hour before he arrived.

Libby Plumley grew up a solitary child who knew the house and gardens intimately. Like Alice Rawe, she can recall the house and surrounds exactly: the back door through which all the postal business was transacted, the red tin roof, the pressed metal walls and ceilings, the huge sash windows in the kitchen, the warm cooking range, the cubby under the tank stand, the swing on the tree, the two orchards, the iron and brass beds, the oak table used for meals, shelling peas and cutting up fruit for jam and sorting letters, the clothesline *that went for about 3 miles down the paddock. I can visualise it so clearly in my own mind, the colours, the curtains that were in the house. Every place and every room I can visualise absolutely, and the garden too, especially my bit of the garden.*[27]

But this was the national capital. Most of the track known as Weetangerra Road was marked for development as the new Belconnen Way. Libby Plumley and her friends, out riding in the late 1950s, were mystified to find surveyors' pegs in improbable remote places. Out the pegs came. When she was fifteen Libby Plumley went to Goulburn to finish her schooling; returning after each term she found pegs thicker on the ground. Canberra was advancing upon Weetangerra like the waters of Lake Eucumbene and the Yallourn open cut. Kerbing, heavy machinery, and then at last new houses began to creep towards the post office along the freshly sealed four-lane highway. Portions were lopped off the 15 ha paddock. The telephone exchange was automated. A post-office opened at the new suburb of Cook, to which Libby Plumley's mother was transferred. Their home was scheduled for demolition and the family was to shift to a new ranger's house 10 km to the west. Everything that would not fit into the new standard government home was shifted to the family farm at Naas. Libby Plumley took down and packed the red sign 'Weetangerra Post Office'.

It was 24 March 1970, and Libby Plumley's twenty-first birthday. Her mother was working at the Cook post office, Libby and her father were helping to move everything out of the Weetangerra home. There would be no more home-produced milk and butter, no home-made melon and lemon jam, no orchard, no chooks, only one or two dogs. As

Libby and her father drove out in the Dodge there were bulldozers roaring in what was left of the home paddock, and half a dozen of their new neighbours digging out plants from what was left of the garden, for their own houses less than 100 m away. Libby Plumley kept her feelings to herself: *We were never a family to sit and express feelings about anything.*

The next day Libby Plumley drove past the post office on the way to work:

I was driving to work at about eight o'clock, and as I was driving past the post office they had just lit it. They had just bulldozed it and poured some kind of igniting fluid on it and up it went. So by the time I got to work I was in tears, I was so distressed at seeing this house go up in flames, it was awful. And I think seeing the old trees that were around the house being singed—it wasn't very pleasant. It wasn't very long before there were houses built there then.

The geographer Yi-Fu Tuan wrote:

> We think of the house as home and place, but enchanted images of the past are evoked not so much by the entire building, which can only be seen, as by its components and furnishings, which can be touched and smelled as well: the attic and the cellar, the fireplace and the bay window, the hidden corners, a stool, a gilded mirror, a chipped shell.[28]

Peter and Sue Boekel have found returning to the site of their former home in Macedon unpleasant, and rarely do so. The block on which their house stood is now a park. Now, as then, they separate the wanted objects from the unwanted. The treasured artefacts are kept in the new family home in Melbourne and brought out for curious visitors: some tiny Delftware clogs, a half-melted ring, a remounted precious stone. Some unwanted artefacts remain in the park, obscure and forgotten by everyone who lived there: a blackened fence post, a painted house brick, a twisted water-pipe. Paul and Eleanor Gray mounted the twists of melted aluminium and the key of the caravan to show to visitors, and keep one or two other retrieved objects in a cupboard. I asked Eleanor Gray if a burnt house should have been left in the town as a memorial. *Oh no, I don't think so. It would be very difficult for some people to cope with. You've got to have a sense that life is reborn.* Anne Boyd discarded many objects from her burnt home at Ferntree Gully but salvaged a few. Most precious was the charred and stained dining table, symbolic of the meeting of like and loving minds, which she initially rejected. *I didn't want to open the shed and see my table and chairs, all smoky and smelling dreadful and black.* It was restored by a friend, but is still marked with the outlines of objects which had rested upon it during the fire. The chairs—now stools because the backs were burnt away—were restored and stand with the table in her rebuilt home. Alice Rawe remembers Shadforth Street not by visiting it but by a reproduction of Tom Roberts's 'Mosman Bay' which hangs over the dining room table, and a photograph of the area from about the same time. Pat Jackson cannot physically revisit the site of her lost home but, from planes, sees the two rectangular concrete slabs on the beach which are all that remain of the changing sheds and of the whole village. Taxiing to the terminus after landing, she tries to estimate where her home stood. Libby Plumley has in her new home the old post-office table, some other furniture and the Weetangerra Post Office sign. Twenty-five years after its destruction we visited the site and found half a dozen pines of the western edge of the windbreak which once surrounded the home paddock. In the front garden of a suburban house, in about the right place, is a scribbly gum with an odd ring of raised knobs in the trunk 1 m from the ground. Was this the tree which stood outside Libby Plumley's bedroom window, a wire guard nailed to it to protect it from the chooks?

So distressed was writer Prue McGoldrick by the loss of her Yallourn home, her town and its social life, that she and her husband determined to take with them what had been most significant in their lives. From the State Electricity Commission they purchased their house, the garage and the Resident Medical Officer's quarters. From the garden they took the Hills hoist, hundreds of bricks, a liquidambar, a portwine magnolia, pot plants and the grapevine. Inside the house, now re-erected overlooking the waters of Lakes Entrance, they placed pictures of the view from their old house, the church and the cinema; to keep the building extensions similar, they hung old Yallourn doors to new frames. Yet, much as Prue McGoldrick loved the old town, almost all they kept was personal and intimate. The McGoldricks did not souvenir street signs or bus stops simply because they were from

Yallourn; they took artefacts of close personal significance. There is almost nothing in their transported home which represents public Yallourn.[29]

There is a vast theoretical literature on the meaning and function of 'home'. Sociologist Doreen Massey suggested that if the social relations of 'home' are more important than the building or the site, then particular identified spaces like 'home' will have a proportion of interrelations which go beyond it. Local places will always have fuzzy edges of movement, communication and social relations.[30] There will always be tension, Massey believed, between the role of consciousness in creating meaning and the role of structural forces in shaping consciousness.[31]

Is it possible to separate the physical place from the people who relate to it and give it meaning? Alice Rawe thought of her Shadforth Street home independently of its people and activities, but found that the actions of her past life had no meaning, had never occurred, except in the context of the actual site. Unlike Andrew Zable, she needed no journey of discovery to make the past real. Her home would always be real to her because of the actions which had occurred within it; even if the house were destroyed the people and their actions would remain real because the site where they had occurred was real, known and understood. Truly the home is the resting place of the meaning of all lost places.

CHAPTER 6 | EMPTY SPACES: THE INUNDATION OF LAKE PEDDER

Until 1972 visitors to Lake Pedder walked 12 km from the nearest road or came by plane from Hobart. Air travellers flew forty minutes westwards into the South-West National Park, around Mount Mueller and Mount Anne, over Scotts Peak, down over the Frankland Ranges and Crumbledown, across the Serpentine River and the Sentinel Range, the dunes, Maria Creek and Mount Solitary, the smaller lakes and the rainforest, and there below was the gleaming ribbon, the focal point of the south-west, the famous beach, into which would have comfortably fitted the city of Sydney from Central Station to Circular Quay. Observers reported that silence was the first impression, immensity the second, then tranquillity, then excitement.

Lake Pedder was inundated in 1972 by the Hydro-Electric Commission of Tasmania. Thousands of people fought to save it, and thousands more tried to reverse the decision after inundation began. Dodgers, posters, clothing, newsletters, books, poems, public meetings, political parties, committees, films, scientific reports, inquiries, stickers, vigils, depositions and petitions were among the measures employed to halt the inundation. All failed.[1] The Tasmanian government felt itself slighted by the mainland. Despite the facts that only a small proportion of the water would be used for power generation, that the power would not be required for more than a few decades, and that the Commonwealth government had offered to pay for a cheaper alternative, the Tasmanian government refused to negotiate. The gates on the Serpentine Dam closed in December 1971. The process which would eventually engulf Lake Pedder and the Serpentine River to form one much larger lake had begun.

Ten months after Lake Pedder rose above its winter level and began to flood the valley, the Whitlam federal government established an inquiry into the circumstances of the inundation. It seemed possible that the inundation might be reversed. Environmentalists believed that though some of the dunes would have collapsed beneath the water the button grass would regenerate, and some of the trees would replace themselves within five years. Might Lake Pedder yet be saved? It was in this atmosphere of desperate hope that Whitlam's 1973 Inquiry, known as the Burton Committee, took evidence in Hobart and Melbourne. For the first and possibly last time in Australia's history, poets and landscape artists were accorded the status of expert witnesses, equivalent to scientists as 'professionals in aesthetics'.[2] The Committee took over 400 pages of verbatim evidence from artists, environmentalists, film-makers, educators, photographers and bushwalkers, and the transcript offers a brilliant insight into the meanings of wild nature to its supporters in the 1970s. Pedder was lost, but the more important point is that Pedder was almost saved, and eleven years later the Gordon-below-Franklin was saved. In 1995 the possibility of draining the Serpentine-Huon impoundment to 'release' Lake Pedder was actively discussed.

No non-Aborigine had ever lived at Lake Pedder. Though a few adventurers like Olegas Truchanas had walked the region for decades, most of the protesters did not know it at all well. Some of the most passionate defenders of Lake Pedder had only been there once or twice. Yet the conservationists' comparative unfamiliarity with the south-west, and their consequent tendency to compare and universalise what many of them saw and felt at Lake Pedder, was not a weakness of the campaign but a strength.

In the tense atmosphere of the Burton inquiry, language faltered. An artist told the Committee that 'visual realities lose by the fact that they have to be translated into language in order to be explained … We have no words for these things'.[3]

Thousands, though, were tried.[4] Some witnesses presented mini-autobiographies: 'I was weeping on that beach and I climbed the Franklands and, looking down, I thought, "I'll never see this again". It began to snow lightly. And I wept.'

Here are a few dozen of the encapsulations:

> A beautiful golden world all of its own, a masterpiece of nature, a manifestation of the Holy Spirit, awe-inspiring, in the presence of something beyond [ourselves], some essential quality of tranquillity, the enormous dynamic of the place, the combination of grandeur and intimacy, the white-man's dreamtime, a resting place, a healing place, a sanctuary, a holy grail feeling, a place of profound beauty, a brilliant rich-blue magnet, of very deep spiritual significance, an intensity of experience, an enchanted shore, the crown jewel, a compelling presence, like coming face to face with Jesus Christ for the first time, of biblical stillness, a profound and awesome silence, the cradle of my adult life, the mystique is not yielded up easily, midnight blue water, a sense of mystery and power, an eerie tug at the throat, enriching and fulfilling, intimacy, a mixture of discovering beauty and a trial of strength, magnificent for its light, its depth of tone, its reflections on so many surfaces, another language beyond the reach of maps or words, to land there on a calm morning was unforgettable.

There were the adjectives: 'unique, intoxicating, superb, beautiful, majestic, world-class, intangible, mystical, primeval, primitive, spiritual, precious, enriching, fulfilling, emotional, unforgettable, incredible, haunting, irresistible, bedazzling, magnificent, exhilarating'.

While the Hydro-Electric Commission regularly described the south-west as useless and unused and the first Lake Pedder as 'modified, disappeared or enlarged', conservationists articulated 'intrusion, destruction, snuffed out, drowned, crucified, tortured, doomed, a national or environmental tragedy, drowned flowers, country that we have ravaged so terribly, short-sighted vandalistic progress' and 'a detestable crime'. The enlarged lake, impounded by an 'evil lump of concrete', was characterised as 'simplicity, monotony, ugliness, of fluctuating levels and dead trees'. What had been 'a water piece of nature' was now a 'reservoir', a 'man-made successor' or an 'artificial man-made pond'.

The witness to communicate her emotions most powerfully was Beverley Dunn, an actor who visited Lake Pedder just once, in 1971, to narrate a film.[5]

> Is it possible to convey accurately and adequately the experience of Lake Pedder? Of course not ... It's like falling in love. Once it's happened, you're not the same person, but it must happen for you to change.

> I was not prepared for the enormous dynamic of the place, the combination of grandeur and intimacy, the silence and the sounds of silence, the gradations and subtlety of colour and the unbelievably vast beach ... The real Lake Pedder is not just a sheet of water, it's a superbly proportioned water in which water, sand, mountains and dunes have been related in a dynamic harmony ... In human terms ... this means that one could walk across the talcum powder and find pale pink sand and if one was looking at the mountain, be unaware that one had started to walk into the Lake. The texture and temperature of the sand seemed to merge for some yards and even after walking for five minutes the water was still only ankle deep, and the colour of the water, influenced by the sand and sky, changed in marvellously subtle ways. With no beach there to refract the light, the colour must be uniform and dark because the button grass water has a golden brown tint. To see it how we did for ten of our twelve days under a clear blue bowl of light, the colour of the water was a combination of cobalt blue and a rich claret paling to the most delicate shades of rose at the large and articulate edges of the Lake where water meets beach ...
>
> Those of us who have stayed there, knew what the experience has done to our inner selves—our souls, our psyches. It was a time one felt privileged to be part of and at one with the universe. Its influence on one's life is intangible, indefinable, but it is there. Must this place of infinite beauty ... be lost to this generation, and of all generations to come?[6]

Only a month before Beverley Dunn spoke her impassioned words to the Burton Committee, a party of conservationists keeping vigil on the sole remaining strip of the dune were striking their last camp. A papier mache bust of Trucanini mounted on a post in the lake bed by the Lake Pedder Action Committee, conflating the destruction of Aborigines with drowned wilderness, was now almost obliterated.

> We stood on the sand. Before us, rustic tables stacked with clean dishes stood above the water, waiting. Waterlogged tents flapped lethargically in the breeze and the green of the trees was fading to brown. The place seemed desolate. Never had the air seemed so full of death. Our throats too knotted to speak, we paddled sadly away towards the shore. As we passed the front of the dune we glanced back. Then we nearly wept. Some of the browned trees which had fallen from the dune's face were sprouting afresh. It seemed a symbol of hope and defiance from these doomed trees.[7]

Though there are passages more elegant, there are few more heartfelt in Australian non-fiction. Yet it is difficult to imagine the residents of inundated towns like Jindabyne or Adaminaby, or destroyed towns like Yallourn and Cribb Island, camping on the edge of desolation to observe and record their own homes and birthplaces disappear forever. Many of those residents found that even to return to these sites was too painful. Why were the Lake Pedder supporters able to do so?

Many of them were younger. They were more articulate and, except for the Yallourn residents, better educated. They had several public forums to reveal their emotions and they knew how to utilise the media. They collectively shared the loss of a single observed place rather than the combined and more devastating collectivity of inhabited home, living area and culture. The world view of wild nature expressed during the campaign to save Lake Pedder was understood by thousands of conservationists world-wide who, while sharing a similar cultural inheritance, had never visited the lake. And there is something beyond that: threatened *uninhabited* space seems to be perceived and experienced by non-Aboriginal Australians differently from threatened *inhabited* place. The reasons why this is so are not clear. Let us try to disentangle these complex matters.

First, what were the meanings ascribed to the drowning valley? The conservationists held as axiomatic the contrast between the corrupting influence of cities and the purity of the natural world. They believed that it was possible for an individual to form an intimate and liberating relationship with wild country, especially if knowledge was acquired through active rather than passive observation. They held that an intimate relationship, or release of the soul through contact with wilderness, was analogous to a religious and transcendental experience, though the mystical relationship was hinted at rather than obvious. They understood that the Lake Pedder region was an artistically indivisible whole. Perhaps most important, it seemed to the conservationists that the unique character of Lake Pedder coexisted with the universal and the abstract properties of all wild nature. *This* wild region was symbolic of *those* threatened wild places world-wide. None of these characteristics was a 'natural' way of finding meaning in wild country. Each was culturally inherited, transformed and shared so widely that Lake Pedder was very nearly saved.

Second, the Lake Pedder conservationists drew their ideas of the significance of wild country from ambiguous historical roots in the European, especially English, aesthetic of what kind of country was to be valued. Wilderness has been perceived differently by different people at different times. Lake Pedder itself was by no means universally admired in the nineteenth century. John Judd, an agricultural and mineral prospector, in the 1850s called the Lake Pedder region an 'almost barren waste'.[8] One of Tasmania's foremost nineteenth-century landscape artists, William Piguenit, probably did not think much of Lake Pedder either; he preferred Lake St Clair,[9] which was flooded in 1937 without much protest.

The historical perception of wild country is equally ambiguous. The ancient Hebrews saw wilderness as formless, lifeless, frightening and unsown—but it was also the place where the Children of Israel came closest to Yahweh.[10] Neither the Greeks nor the Romans thought much of wild places, and later Europeans inherited the literary imaginings of Pan and the satyrs, or filled their own wildernesses with monsters like Beowulf. In mediaeval times, wild tribes were supposed to inhabit fearsome wastelands, but monasteries were removed there for purity and refuge. Puritans fled from corruption to the forests, but American pioneers saw a significant purpose in taming the wilderness.[11]

From the mid eighteenth century the perceptions became less ambiguous and more recognisable. This period, in England, is a convenient departure point. To locate the perceptions of the Tasmanian wilderness in the 1970s, let us trace three developments, the first artistic, the second historical, the third philosophical.

THE MEANING OF WILDERNESS I: CONTROLLING THE LANDSCAPE, OR BEING CONTROLLED

By 1770 observers had learnt to separate the abstract, generalised vision of landscape (rolling hills, lakesides, open parklands) from the actual intellectual, emotional and historical associations connected to the scene they were looking at. According to the dictates of artistic sensibility, the cultivated observer should be able to contemplate a landscape as if it were a picture, and move about until the most aesthetically satisfying vantage point was found. Often that was an elevation, from which the elements composed themselves into a regular and comprehensible pattern. Imposing natural features such as mountains were mostly avoided by artists and poets while more gentle natural phenomena were surveyed and engaged by the eye but, in the end, controlled by the head. Each geographical feature was depicted as known, understood and in harmony with other elements. Cultural historian John Barrell compared Thomas Gray in a country churchyard, alone, melancholy and philosophic—but controlled within the poem with Thomas Gainsborough's painting of two courting figures beside an ancient oak, gnarled and blasted—but controlled within the picture.[12] Nature, according to convention, challenged the observing artist, but the artist overcame nature through the act of observation and conceptual enclosure. The elements of nature thus arranged in a style recognisably picturesque were now able to be read and could be classified, an individual example to a species to a genus to a class.[13] In this sense, though the artist or poet chose an individualised perception or vantage point, the country was unimportant compared to the *idea* of landscape to be imposed. Artists and writers were not yet participants emotionally involved in the scenes they confronted.[14] Not so at Lake Pedder, where most observers felt awed or dwarfed by the vast panorama.

Observers became more emotionally involved after cultural critics of the later eighteenth century extended the elements of nature that could be artistically observed. In 1794 Uvedale Price published his *Essay on the Picturesque,* in which he argued that there was more to human feelings than the classical categories of the Beautiful and the Sublime. A third was the Picturesque; landscape worthy of admiration might include roughness in texture, irregularity, asymmetry, partial concealment, the unexpected and natural occurrence rather than artificial contrivance and sudden contrasts.[15]

Wordsworth, younger than Uvedale Price and much concerned with the ferment of ideas of the Enlightenment, thought that this picturesque style of apprehending nature, superior as it may have been to the classical forms, encouraged superficial judgments which interrupted deeper feelings and a state 'in which the eye was master of the heart'.[16] Why should the heart not be master? The intellectual roots of that answer lay in Primitivism, which was a reaction to the intellectualising and rationalising European Enlightenment. One of the tenets of Primitivism was that the further humanity journeyed, physically or philosophically, from nature towards civilisation, the more its unhappiness increased. Byron wrote:

> There is a pleasure in the pathless woods,
> There is a rapture on the lonely shore.
> There is society where none intrudes …
> I love not man the less, but nature more.[17]

This change in perception has been described as from a natural world capable of rearrangement and control to a natural world demonstrating the principle of creation. Even as Uvedale Price enunciated the contemplative principles of the picturesque, artists were ceasing to empty landscape of mystery by enclosing or controlling it. A generation after the publication of Gray's 'Elegy', artists were expected, or expected themselves, to be awed or overwhelmed by nature.

THE MEANING OF WILDERNESS II: DRAINING PARTICULARITY

Why was that? Cultural historian Raymond Williams argued that the eighteenth-century English landscape, changing faster than in any previous period, was most regretted by those who saw the features of their own familiar country disappearing. Williams nominated nostalgia for vanishing familiarities as the instigator of changing taste.[18] John Barrell also believed that the pace of change altered the way in which rural scenery was valued. The process of English landlords enclosing their agricultural lands for sheep pastures had for centuries rendered thousands of people homeless and hundreds of villages deserted;[19] in the eighteenth century so widespread had enclosures become that the changes in the southern English landscape were apparent even to strangers. Footpaths vanished before new drainage systems, allotments and fields became rectangular and uniform. The old farm tracks which for centuries had led those who knew the way to village, church and tavern, were replaced by inflexible trunk and turnpike roads which followed land contours and required neither local knowledge nor affection. The turnpikes were linear, linking supply village to market town, rather than spiral, linking outfield to village. The major movement directions joined, not circled, dots on a map. The new roads went not *to* but *through*, the village church ceased to be a local centre and the departing populace, as Goldsmith observed, took the new straight road to emigration:

> Even now the devastation is begun,
> And half the business of destruction done;
> Even now, methinks, as pondering here I stand,
> I see the rural virtues leave the land.[20]

Such highways were also used by a rising class of rural professionals—surveyors, valuers, brokers and auditors—who throughout the eighteenth century became much more numerous and mobile. Their task was to comprehend, to mentally encompass a landscape assumed to be comprehensible, and the fastest way for newcomers to comprehend unknown country was to destroy the local knowledges, to discard old names, to change the geographical irregularities into recognisable templates, and generally to appropriate the unknown by means of the turnpike, the rectangular fence and the comprehensive view. When not engaged in calculation, surveying or map-making, the professional traveller or government employee was on the way to a new district. No location was any longer a place-in-itself, but was mediated by other places. Each locality was part of a larger area. What was special about any particular place, Barrell remarked, was forgotten or uninteresting; the map was drawn blank by the surveyor.

The casualties of this profound change were the local knowledges and particularities. Enclosure of open fields obliterated some villages, and re-aligned others which survived. Local people went to other towns, cities and countries and took with them that intense local knowledge which evolves and flourishes through ignorance of other places:

> The hostile and mysterious road-system was tamed and made unmysterious by being destroyed; the minute and intricate divisions between lands, strips, furlongs, and fields simply ceased to exist ... Everything about the place, in fact, which made it precisely *this* place, and not that one, was forgotten.[21]

Not everyone agreed that the changes were for the worse. Every generation has its developers, and rural writers like William Cobbett and

Arthur Young thought enclosed fields beautiful and common land ugly.[22] But for good or ill, travellers and writers and surveyors and artists were beginning to compare one place with another. Even William Wordsworth, meditating upon unique locations, separated the objects in view from wider philosophical reflections upon those objects. Whether commercial or poetic, general reflections based upon the sum of many particularities were becoming a dominant mode of English thought by the beginning of the nineteenth century. People who had never been to particular locations heard or read about them—and understood.

THE MEANING OF WILDERNESS III: THE VIEWER AND THE VIEWED

The third element in the formation of the 'Pedder' view of wild country was the consciousness of the distinctiveness of self. A critical point in the history of modern European thought is held to be the moment in the eighteenth century when the viewer and the view became separated. Landscapes had been admired before, but from that point, some 200 years ago, an observer might look at a landscape and be conscious that he or she was doing so.[23] Lookouts had not always been places where one stopped for lunch, breathed deeply or set up the easel. A 'shrewd and sensible' woman of the English Lake District told Wordsworth, 'Bless me! Folk are always talking about prospects [views]; when I was young there was never such a thing neamed'.[24] This sense of an individual psyche, its place in the world and its consciousness of the passing of time, was not invented by the Romantic observers of nature; the phenomenon had been evolving since at least the time of Locke. But Romantic observers like Wordsworth and Coleridge lifted the sense of individual consciousness from the realm of aesthetics to the key philosophical relationship which until very recently dominated the discourse of Europeans about wild country. 'What is the relationship between this country and me?' Wordsworth's particular contribution was reverie, the 'imaginative melting of man into outer nature' coupled with a sense of an irrevocable human past and a continuing and present landscape, from which nineteenth-century poetry drew some of its greatest power.[25] The innocence of humanity, corrupted by civilisation, could be regained by an intense communion with nature, by individuals achieving their own relationship with wild country.

The semi-mystical relationship between individuals and the Lake Pedder wilderness begins to coalesce. While mountains to one seventeenth-century observer were 'Nature's Shames and Ills', by 1800 they were becoming 'temples of Nature built by the Almighty', 'natural cathedrals' and 'natural altars'. The city and the wilderness had become philosophical polarities. Mountains transcended trivialising rationality. Artistic sensibility used its experience to organise its exploration of the natural world, and related phenomenon to phenomenon.[26] Nature, not organised religion, might be sufficient for humanity to fully experience emotions that were felt rather than observed. Byron's conception of an intimately personal, and bilateral, relationship with sublime wild country was at the heart of the Lake Pedder protest.

In the settler societies wilderness acquired more specific aspects. Americans, anxious that their country could not match European ruins and spectacles, surpassed Wordsworth in the solemnity of their contemplation of nature and, in the opinion of geographer David Lowenthal, consciously substituted a wilderness aesthetic for the European ruined abbeys and castles which they did not possess.[27] The great American exponent of the morally rejuvenative powers of wilderness was Thoreau:

> Is it some influence, as a vapor which exhales from the ground, or something in the gales which blow there, or in all things there brought together agreeably to my spirit? ... The trees must not be too numerous, nor the hills too near, bounding the view, nor the soil too rich, attracting the attention to the earth. It must simply be the way and the life—a way that was never known to be repaired nor need repair within the memory of the oldest inhabitant ... There I can walk, and recover the lost child that I am.[28]

Awe at the might of wild nature reached its endpoint in America in the philosophical state of transcendentalism. To Thoreau, transcendentalism was a higher reality than the physical world. The linking point, or correspondence, with that higher reality was the physical world itself. The importance of the natural world lay in its reflection of universal truths which humanity could share only through intuition and imagination. A hundred years before the Burton Inquiry into Lake Pedder, American transcendentalist John Muir wrote of nature as 'a window opening into heaven a mirror reflecting the Creator. Presently you lost consciousness of your separate existence, you blend with the landscape and became part and parcel of nature'.[29] The transcendentalist experience seemed particularly likely to occur in mountains, the wilder the better. Climbing mountains in Canada was said to be 'coming face to face with Infinitude'.[30]

Intellectual and physical changes were altering conceptions of nature. The theory of evolution eroded the idea of humanity at the apex of the natural world and allowed observers to think of themselves as equivalent to other forms of biological life. Terms like 'kinship' or 'companionship' with nature became common terms to Canadians in the 1880s. Poems were written about what it felt like to be an animal or a flower. The Canadian poet Lampman wrote, 'We know with the fullest intensity of sympathy that we have one berth with everything about us, brethren to the trees and kin to the grass that now ... flings the dew about our feet.'[31] The famous Australian naturalist Myles Dunphy believed that 'It is paradoxical that wilderness or primitive bushland should be one of the really indispensable necessities of modern existence in its soundest sense. This is the new and modern view. Where else can man go to escape his civilisation?'[32]

This was the legacy of a century of involvement with wild nature. Most readily recognisable at Lake Pedder was what Raymond Williams called the lonely creative imagination, the observer driven back from the cold world, seeking to find or recreate humanity, projecting his or her personalised feelings into the landscape, a 'subjectively particularised and objectively generalised Nature'.[33] Space had become the common symbol of freedom in the western world. Space suggested the future, it suggested action.[34] The losers in this enormous shift were those who valued the particular rather than the general, the known rather than the unknown, the lived town or city rather than the deserted wilderness.

Deepening interest in the general and lessening interest in the particular had entered the psyches of travelling Europeans and settler societies, and in the nineteenth and twentieth centuries became the dominant world view by which some places came to be valued and others devalued. The nineteenth-century poets, artists and writers generalised their own place to make it relevant and knowable to others. Wordsworth himself, as John Barrell remarked, was always on the turnpike to somewhere else.[35] His poetically described landscapes were a means of preserving memories and of unrolling reflections on philosophic universals. His relationship with wilderness and, 150 years later, that of the Pedder protesters, was dialogic. Wilderness promised the interrelation of humanity and the cosmos. The spirit of a place, be it at Tintern Abbey or Lake Pedder, was found by looking through it as well as at it. The most meaningful particular wild country represented the notion of wild country generally.

AUSTRALIAN WILD COUNTRY

What did Australian colonists do with their own wild country? Lexicographer Bill Ramson considered the processes by which the invaders tried to establish a personal identity in what was (to them) empty and unknown. By examining the words used to describe the Australian environment at different periods, Ramson demonstrated that the earliest Australian-European vocabularies were utilitarian rather than interpretative. There was a tension between resemblance and difference in the first European-Australian vocabularies. Unfamiliar plants received familar names of 'wild fig' or 'wild geranium' and in the mid century, phrases like 'the far north', 'far out' and 'the never-never' demonstrated a continuing apprehension about the vastness of the land. The great era of possession—exemplified by the number of new Australian possessive words or phrases entering common use—he found to be the third-half century of occupation, 1888–1938. Expressions entering the language in that period, like 'the Red Centre', 'the Great Outback', 'closer settlement', 'the man on the land', and 'wheat cocky', demonstrated to Ramson the colonists' hard-won familiarity with the landscape and their self-identity within it.[36] The same familiarity has been observed in some of the paintings of landscape artist von Guérard. His rural scenes often included cleared and farmed country. By placing the homestead and its improvements in the distance in certain paintings, von Guérard apparently celebrated the absorption of humanity into the landscape.[37] The Australian landscape was becoming known, understood and psychologically enclosed.

Psychological understanding was only one strand of possession. Following is explorer David Burn's initial reaction to wilderness in the mountains of Tasmania.

> It is beyond the power of any pen it far surpasses mine to convey the faintest idea of the magnificent grandeur—the boundless variety—the romantic wildness—the pictorial loveliness—the enchanting grace-fullness of the stupendous panoramic scene we here beheld ... the rich and the luxuriant—the savage and the bleak—the sublime and the picturesque ... How shall I feebly attempt to paint the charming plains beneath? broken into countless varieties of hill and dale, or floral mead or grassy knoll, of verdant copse and sunny bank—glowing, in fine, with every attraction of pictorial beauty or romantic fascination, they must be conceived—description fails me.[38]

As knowledge and experience of the Australian bush expanded, the sense of fear in wild country took more conventionalised literary forms. The transformation was exemplified by Rosa Praed's fictional creation Lady Bridget, who in her new home in remote Western Australia is crushed by the 'weirdness' of her 'mournful' environment.

> She ate her solitary dinner and paced the verandah till darkness fell and the haunted loneliness became an almost unbearable oppression. Vast plains, distant ranges, gidea scrub and the far horizon melted into an illimitable shadow. The world seemed boundless as the starry sky—and yet she was in prison.[39]

Such is the response of the metaphorical outsider who remains at a distance, emotionally involved in the landscape yet not at one with it. Many nineteenth-century Australian observers might name or rename geographical features and admire a natural world capable of abstraction and comparison, but it was also a natural world which retained an unfathomable remoteness.

Art historian Tim Bonyhady has traced the early development of the aesthetic of the gloomy and weird in the Australian imagination. Conrad Martens, for example, painted some of his landscapes as hostile and forbidding, containing few or no human figures.[40] But the high point of the convention of awesome melancholy was achieved not through painting but literature. 'Weird Melancholy' became for a time the theme of Australian artistic culture, especially after the publication of Marcus Clarke's 1876 introduction to an edition of the poetry of Adam Lindsay Gordon.[41] Clarke needed only seven lines of lugubrious prose to describe cockatoos 'shrieking like evil souls', kangaroos hopping 'noiselessly', the 'rustle of strips of bark', 'horrible peals of laughter', 'melancholy gums', 'frowning hills', and sundry other objects 'grotesque', 'ghostly', 'silent', 'dismal', 'fear-inspiring', or 'gloomy'. For two decades the themes of eternal solitude and gloomy exclusion from civilised life were seldom far from painted landscapes until put to flight by the plein air school of Heidelberg painters.[42]

Though the theme of weirdness remained subdued in painting, it was an enduring element in literature. Henry Lawson's bush was 'weird' as well as 'the home of the weird and of much as is different from things in other lands'.[43] Novelist Eleanor Dark portrayed Arthur Phillip, newly arrived from England, staring from his flagship at the shores of Port Jackson:

> No hostility from man or beast. Nothing to grapple with. No call to action. Only this darkness which kept the senses painfully alert, this silence which tautened the nerves and this devilish illusion of arrested time. 'Tomorrow' he said briskly to himself, but the word was empty. The ageless land had drained it of its meaning and its promise. There was nothing but oneself, a tiny spark of consciousness, alone and aghast in this unconquerable silence.[44]

Bruce Chatwin's *The Songlines*, published in 1971, contains this exchange:

> 'It's a weird country', I said.
> 'It is.'
> 'Weirder than America.'
> 'Much', he agreed. 'America's young. Young ... and cruel. But this country's old. Old rock. That's the difference. Old, weary and wise.'[45]

The most graphic writer in the idiom of weirdness was D.H. Lawrence:

> Yet, in the wild bush, God seemed another God. God seemed absolutely another God, vaster, more calm and more deeply, sensually potent. And this was a profound satisfaction. To find another, more terrible, but also more deeply-fulfilling God stirring subtly in the uncontaminated air about one. A dread God. But a great God, greater than any known. The sense of greatness, vastness and newness, in the air. And the strange, dusky, grey eucalyptus-smelling sense of depth, strange depth in the air, as of a great deep well of potency, which life had not yet tapped.[46]

We are not far short of the mysticism implied by the Pedder conservationists, and we are drawn closer still in Rosa Praed's novel. Lady Bridget's husband-to-be, the bushman Colin, is not imprisoned but liberated by the darkness and silence of the bush:

> I'd like you to camp out in the Bush sometime, Lady Bridget, right away from everything—it'd be an experience that 'ud live with you all your life—My word! It's like nothing else—lying straight under the Southern Cross and watching its pointers, and, one by one, the stars coming up above the gum trees—and the queer wild smell of the gums and the loneliness of it all—not a sound until the birds begin at dawn but the hop-hop of the Wallabies ... truly, it's the nearest touch with the Infinite I've ever known.[47]

Most of the mystical and semi-mystical passages about Australian landscape stop at this point: transcendentalism has never found much favour in Australian non-fiction. Non-Aboriginal Australians commonly feel themselves receiving the spirit of a site, but they seldom claim to become spiritually one with the site itself. They have seldom felt a particular 'spirit of place' as a bridge to a mystic transformation of the self into divine understanding. The Pedder conservationists reflected that tradition. Like Lawrence, most conservationists found the bush profound, silent, potentially transforming, full of presence. And there, like Praed's Colin at 'the nearest touch with the Infinite', they stopped. It was far more common to *liken* one's experience at Lake Pedder to a mystical transformation than to claim to have been mystically transformed. Weirdness and melancholy, which we might describe as a laconic Australian equivalent of American transcendentalism, has a place in the discourse of Australian wilderness as a touchstone to relate oneself to the cosmos. In 1973 Marcus Clarke's 'frowning hills' had become 'the profound and awesome quiet' imbibed at Lake Pedder, his 'melancholy gums' were now 'the breath of a lone seagull coming in to land at this island haven'. Both aesthetics presuppose that wilderness is larger than everyday urban life; both presuppose that humanity is diminished before a vaster presence.

Weirdness takes a little of the terror of the first colonists, the non-residents' fear of the unknown, the elation of the sublime and the almost-transcendentalism of the later Romantics into the aesthetic of wilderness recognisable in modern Australia. Beverley Dunn expressed all these elements to the Burton Committee:

> It's on a scale comparable to a Sibelius symphony or a Greek tragedy or a religious experience. Initially the grandeur was overwhelming, but after one had been at Pedder for a few days the scale of this grandeur became relevant to the great eastern beach on which one could stand, run, walk or just be and feel privileged to be just part of that magnificent wilderness and somehow part of the universe [universal] spirituality.[48]

Lake Pedder was particular, but was also universal. So were Lawrence's bush and Colin's starry sky. As a literary critic remarked about the subjects of Romantic artistry, 'The specific is not rejected; it is loved for itself. But the specific is connected to large rhythms.'[49]

SAVING THREATENED PLACES: PARTICULAR OR GENERAL?

This connection of the specific to the larger rhythms is what, if anything, will re-create Lake Pedder, for the aesthetic has been translated from literature into law. Since the 1980s, most special places threatened with destruction must, to be saved, be capable of being universalised. Three of the four criteria for the nomination of natural environment to UNESCO's World Heritage List include the word 'example'. The first criterion is: '[Nominated property should be] outstanding examples representing the major stages of the earth's evolutionary history.' Only the fourth, concerning the preservation of threatened species, privileges the particular as valuable, and even these species have to be 'of outstanding universal value'.[50]

Particularity is also unwanted in the preservation of the built environment. A place must not just be special to somebody, it must either be representative or a fine example. Mere uniqueness is insufficient. Like the criteria for saving wilderness, cultural property preserved under the UNESCO criteria must be, for example, 'a masterpiece of the creative genius' or 'an outstanding example of a traditional human settlement which is representative of a culture'.[51] In neither the canons of Romantic poetry nor in the environmental bureaucracy is particularity respected for its own sake.

That is the cost. Outsiders—the decision-makers—no longer understand specific localities in relation to their specific meanings. Throughout most of the twentieth century changes analogous to the English enclosure movement have steadily emptied a thousand Australian localities of their special significance. Our villages had developed less organically than their English equivalents. Except perhaps in the far north and west, a sense of a greater Australia linked site to region, and the particularity of individual towns probably began to decline as they were joined by the railway. In the 1920s a wheat-growing town could be sacrificed by state government if there was another town from which wheat could be railed more cheaply. In the 1950s the motor car and sealed road were penetrating everywhere. Lessening employment in the 1970s and failing local economies in the 1980s accelerated the devaluation by outsiders of villages and towns held special and worthy of preservation only by their own inhabitants. As the poet Sally Roberts ironically described an inundated Welsh village:

> Nothing's gone that matters—a dozen farms,
> A hollow of no great beauty, scabby sheep,
> A gloomy Bethel and a field where sleep
> A few dead peasants.[52]

How could Yallourn, Cribb Island, Adaminaby or Jindabyne survive under the crushing weight of such outside valuation?

Even large inland cities were affected. Goulburn, for example, a fine city in southern New South Wales, was built to face north-west into the pastoral and agricultural regions which it serviced. That trade declined while the road traffic to Melbourne, and later to Canberra, increased. From the 1920s cars and trucks thundered through a back laneway becoming known as the Hume Highway. Goulburn had to turn around to face the east. Shops were built along the suddenly developing highway. The city boomed—now in a different direction—as a stopping place between the capital cities. Then in 1993 the superhighway, like an eighteenth-century turnpike, bypassed the city altogether. Apart from residents and regional tourists, Goulburn was now just a sign on the main road to somewhere more important. Goulburn ceased to matter.

Except to the people of Goulburn. The particularity of a thousand loved Australian places survived among those to whom they were most familar. Talbingo, like Adaminaby and Jindabyne, was flooded by the Snowy Mountains Authority, but it did not sink unlamented. There were particularities of Talbingo known only to those who loved it:

Jack Bridle's Farewell to Talbingo

The stream where Hume and Hovell crossed about 1822
Is now more like the Lachlan or the Bogan at Dandaloo,
Old Brander's Flat, may she rest in peace, where we swam so long ago,
Janey's sands and boulders too, it hurts to see them go.
Where we tramped along the grassy bank for many a happy mile,
With friendly old Talbingo watching o'er us, all the while.

...

For the water is creeping over where once were fields of clover
And on the flats the weaving yellow corn
Now I hear a fiendish rattle where once were grazing cattle
Oh how I loved the land where I was born.[53]

In these verses from 'Adaminaby, The Old Town', the particularity of special places like Talbingo's Brander's Flat and Janey's Sands is matched by the families known only to Adaminaby residents:

> Then talk flowed thick from Joe and Dick
> of Kennedys and Shanleys
> And Yens as well. You almost know they'll have to go
> The Mackay men, and the O'Neills will sell.
>
> They'll all move out without a doubt
> Wave goodbye to Stewart and Rossiter
> to Norman Lett, he'll go I bet
> When the old town's drowned and they transfer.[54]

Such raw intense responses were incomprehensible to most outsiders. Probably the towns always were doomed. Only the professional universalisers, the poets and film-makers, understood the potential of enlarging and relating local particularities to those of all humanity. In directing *A Town to be Drowned*, Robert Raymond understood the death of Adaminaby beyond its merely local pathos because he had witnessed the inundation of villages in Africa. Douglas Stewart urged readers to mourn the drowning Jindabyne in universal, not local, human values:

> Finally for the mystery and the pathos
> That seep from earth and bubble out from water
> In any place where men have lived and bred
> And feuded with each other.[55]

The poet Stephen Edgar linked 'threatened [rational] senses' to universalising intuition through the shadows on the drowned beach and the wrested myrtles of dead Pedder:

> **Pedder**
>
> The shadow in the surf that empties beaches;
> The drowned bell in a western bay
> Clanging on the hinge of a grey
> Tide: It is something in water that invests
> Power in distortion, teaches
> The threatened senses it arrests
> What things in thin air are too frail to say.
> The banners of a slaughter-
> Wasted army heave in the slow lake water
>
> More largely than in winds that threw them over.
> A launch above the flooded lake
> Extends the feather of its wake,
> Brushing peaks. This is vertigo. I ride
> Where the sky was and cloud cover.
> Below, like air, the currents slide
> Down hills and slur through filmy trees to make
> As clouds would on the sand
> Viscous shadows where the wrested myrtles stand.[56]

Lake Pedder liberated the senses; it was 'any place' where nature was held to have wrought a masterpiece. It was loved as a supreme example of universal wilderness values as well as an overpowering individual presence. That is why it was nearly saved; the protesters almost succeeded because Pedder was understood and fought for by those who conceived it as a symbol as well as a precious object. This was no rustic town of intimate human association, quaint individuals and old families. Almost all the non-Aboriginal names of the landforms surrounding Pedder were conferred by nineteenth-century explorers. Very few place names commemorated a memorable incident, for almost none had occurred. Conservationists world-wide connected this particularly beautiful natural phenomenon to the generalised aesthetic of wild country. A popular song during the 1980s campaign to save the Gordon-below-Franklin linked that particular Tasmanian valley to the universal wilderness aesthetic twice in one verse:

> Let the Franklin flow, let the wild lands be
> The wilderness should be strong and free
> From Kutakina to the south-west shore
> The wilderness is worth fighting for.[57]

Universalism is a successful tactic as well as a philosophy. The Franklin River flows as free as the day the song was penned; Adaminaby and Talbingo are lost forever.

Cultural historian Bernard Smith has argued that the predominance of Australian landscape painting over other subjects was responsible for creating that consciousness of Australia which drew its emotional iconography from the bush. We can see now that there were political advantages in such an iconography: one was the aesthetic discourse of wild country which saved the Franklin River. Smith noticed several disadvantages. Since most Australians lived in suburbs, the iconography was a false consciousness; it disguised an old European Romanticism and encouraged the belief that Australians were distrustful of theory, if they were not altogether mindless.[58] There was a further disadvantage unnoticed by Smith. The elevation of the undifferentiated and unlocalised bush to iconography made it correspondingly harder to save from destruction the local, the familiar, the specific, the lived-in, the un-unique and the un-universal—especially the suburb, the street, the private house and the country town.

Compared to the rich and complex aesthetic of wild country bequeathed to us by nineteenth-century Romanticism, Australians have no adequate discourse to conceive, describe and hence defend our apparently ordinary homes and suburbs from speculators and freeway builders. We have the words and feelings but not the rationalist context into which our expressions of meaning can be understood by planners and assessors. That is one reason why, compared to wild country, they fall before the developers' bulldozers so easily. The next chapter describes a rare instance of an outsider failing to impose its will, on the urban community of Darwin.

The Burton Committee took 400 pages of evidence on the question 'What does Lake Pedder mean to you?' In the 1990s, neither developers nor environmental impact assessors are required to ask that question of people trying to defend their suburbs and streets. Lake Pedder may yet be drained, and the lesson of the long campaign appears to be: those who wish to save their urban living area from destruction should seek to generalise and universalise its value.

CHAPTER 7 | DARWIN REBUILT

The condensation heat energy released by a tropical cyclone in a day can supply the electrical needs of the United States for six months. The same energy released upon Darwin during the night of 24–25 December 1974 flattened the city. Darwin is rebuilt—unlike Adaminaby—on the same site, but it is not the same city. Former residents who returned for the twentieth anniversary commemorations in 1994 found much of the city and suburbs unrecognisable. Yet while the Pedder conservationists failed to save their lake from destruction, Darwin residents preserved their city, not from the cyclone but from the uniformalising plans of city developers. Darwin's rebuilding took very much the forms which its citizens wanted.

Residents of pre-cyclone Darwin recall the easy pace of life, the absence of traffic lights and rush-hour tensions, the toleration of ethnic differences, the social life in gardens or under the stilts of the raised houses. In the 1960s it was possible to walk up the main thoroughfare and know almost everyone by name or face. In old Darwin, according to trade union official Curly Nixon, 'you never got an invitation to a party but you got abused if you didn't go'.[1] The less romantic recall rougher aspects. Botanist George Brown found Darwin to be a male-dominated and hard-drinking town. He had his first fight soon after he arrived in the mid 1960s when he walked into Fannie Bay Hotel 'where a bloke walked up to me and said "Can you fight?" And I said, "Well I can but I don't want to." And he said, "Well you're gonna".

And he landed one on me and I had to and that was part of the entertainment.'[2]

In the 1960s Darwin was a stopover on the international air route to Europe. In the early 1970s the city was a step on the hippy trail to India and South-East Asia, and at any moment contained several hundred transients. They were a part of Darwin which was already changing before the cyclone. Another change was the increasing number of public servants occupying the new treeless suburbs of Parap, Nightcliff, Wagaman, Anula and Alawa. Popular sentiment held them to be uncommitted and isolated, but it was clear that in the early 1970s more families were coming to settle permanently in the 'top end'. One permanent resident was Harry Singh, a Kiyuk Aborigine, who was employed erecting houses at Delissaville (now Belyuen) on the other side of Darwin Harbour.[3] A new arrival was Sally Roberts who liked the lifestyle but found the northern suburbs hot and noisy: 'One could hear almost everything from one elevated house to the next.'[4] Another new arrival was the Hancock family. Peter and Kass Hancock moved to Wagaman, the last suburb to the north-east of pre-cyclone Darwin, in 1972. They too liked the easy pace of life and the way that even strangers said hello in the streets. They enjoyed camping on weekends, the openness, the young people. Peter Hancock was employed by the Department of Works and ran a small manufacturing business; Kass taught pottery at Wagaman primary school and worked as a marriage counsellor. Life for the Hancocks was energetic and full. In Wagaman young families were busy setting their new houses in order, making friends and setting out gardens.[5]

Though there were plenty of warnings, the cyclone predicted to arrive on Christmas Eve 1974 was not taken as seriously by Darwin citizens as it might have been. As at Macedon, a minor alarm a few weeks before the disaster caused some overconfidence. Nevertheless most people seem to have taken the elementary precautions of taping windows, turning tables upside down, locking up or anchoring loose objects, filling the bath with water and opening the windows to leeward of the northerly wind and rain which was steadily gathering strength from about 6 p.m.

The housing workers at Delissaville had knocked off for Christmas and were drinking in the Mandorah pub when the manager of the housing association realised that the warnings were no false alarm. He recalled Harry Singh and other workers at about 6 p.m. *We were watching the seas, the stirring up of the seas, with the wind whistling and listening on the radio at the same time.*

By 11 p.m. it had become dangerous to drive. Sheets of iron were whipped from roofs and, spearing across roads, menaced drivers anxiously returning to their homes. Locked louvre windows ceased to offer protection from the driving rain spraying north-facing rooms with fine mist and collecting in puddles on the floor. By midnight television aerials had blown away, radio transmission had ceased and the power had failed. A family opened their Christmas presents at a minute after midnight thinking that it would be their only chance. Midnight mass at St Christopher's Cathedral was brought to a premature halt as the wind whistled through the rafters and alarming crashes from outside echoed through the church. By 12.30 a.m. the sound of the wind had risen from what one listener described as a 'squealing roar' to 'white noise'.[6] Some families retreated from front rooms into bedrooms to shelter under the largest bed as their homes begin to disintegrate around them. Others made for the bathroom and braced themselves against the door.

Harry Singh went to sleep in his demountable Delissaville house. He was woken two hours later to find *I had no roof and one side of the wall of the demountable had gone. The house kept lifting up and down—it was on two foot [60 cm] piers—and it was being lifted up for about a foot and then dropped down, up and dropped down each time.* He headed for the toilet and made a nest of blankets for himself and his adopted son. At about 2 a.m. Peter and Kass Hancock and their three children, sheltering in the bathroom, heard their house breaking up around them. The roof collapsed, virtually imprisoning them, but their lives were saved by the piano on the other side of the wall, which prevented the last of the roof falling in and what remained of their house from being blown away altogether.

Some time between 2 and 3 a.m. the noise of the cyclone abruptly ceased as its eye passed over the northern suburbs. A few incautious residents went outside to survey the damage and swap beers and stories. The more wary knew the cyclone would return in a matter of minutes and debated whether to remain where they were, take shelter in a car or drive to a safer location.

Some people saw visions of environmental ferocity that very few have lived to recount. Harry Singh recalled the red, yellow and purple lightning. *Every time the flashes occurred you could see bits of branches and timber and iron and whatever flying past and overhead.* An Australian Broadcasting Corporation employee, Donald Sanders, remembered:

> looking out the back door [of the ABC building] and seeing this huge chunk of roofing material, corrugated iron and cross pieces of 2 by 4 timber, flying through the air from an adjacent police boys club. It sliced through the side walls of the TAA [now Qantas] building which abutted us at the rear and just disappeared inside and then again almost in slow motion the roof partly sagged and then just as the wind entered the building through the broken walls, it just literally went up as if it had been hit with a 250 lb bomb, in slow motion. Just very gracefully sort of up, went its various ways, splintered up, disintegrated and then collapsed in a heap in the middle again.[7]

Probably the greatest damage to Darwin was sustained after the eye crossed the city, though it is not clear whether the winds were stronger or whether the houses, already substantially damaged, were more vulnerable to a storm from the opposite direction. The coconut trees along Gilruth Drive were not broken but neatly twisted clockwise out of their root structure, all pointing northwards towards Fannie Bay. Survivors emerged to find that houses which had occupied the other side of the street and beyond no longer existed. In the northern suburbs dazed residents could see the sea formerly concealed by 500 m of suburbs. Only 408 out of 10 000 Darwin homes were intact, and in the newer treeless suburbs the destruction rate was 100%. Each survivor's story seemed more amazing than the last. A woman sheltering on her kitchen floor was blown bodily away with the remains of her house and found herself on the lawn with the refrigerator beside her. A family emerged from the wreckage to find the dog asleep on the Christmas presents. There were accounts of appalling deaths: a woman sheltering in the bath under two matresses was speared by a piece of wood which pierced the roof, the mattresses, the woman, the bath and the floor beneath. A sailor on duty returned to find his wife and two children crushed and asphyxiated beneath the collapsed double bed.[8] Thirteen children were known to have died out of a land death total of forty-six, sixteen people were missing at sea. The number of dead and missing crept up. Darwin was a dead city. Suzanne Parry was holidaying in

Perth when she heard the news: like Margaret Johnson she wept not just for her home or the people, but for the whole city, that loved place.[9]

Like the residents of Macedon, shocked and distressed survivors were surprised to discover that others had survived. Harry Singh found a contractor's caravan jammed against some trees 15 m from where it had been parked. Its occupant was still inside, *sort of crouched down ... couldn't talk, his eyes wide open and mouth wide open, just crouched down*. Peter and Kass Hancock were ecstatic that they and their children had survived. They recall that it was such a joy to hear that friends had survived, such an anxiety that others were injured that news of the dead was filed away in the memory. Grief could come later. New arrivals were shocked almost speechless at the abrupt and total physical destruction. Ruary Bucknell stood on Boxing Day at the place where his Wagaman home had been.

I sat down at the house and I think I bawled my eyes out for about—I don't know—ten minutes or so. Just couldn't get over the shock of the whole thing. Not so much our house ... but the complete state of devastation. I don't think it's anything that a normal person can comprehend.[10]

All that remained on the raised floor of his house was part of the kitchen, the stereo upside down full of water and plaster, and a cabinet, the contents miraculously preserved. The wardrobes had vanished, but lying on the lawn was a camphorwood chest still filled with blankets. The linen press was by the front fence, and some of the bed base, torn and ripped and soggy, was in a neighbour's yard. Ruary Bucknell's wife had asked him particularly to search for her jewellery box. The box had vanished but under a piece of galvanised iron were half a dozen of its precious items and a plastic doll.[11] A fortnight later a public servant toured the city and wrote in words similar to those which the *Cooma Monaro Express* reporter had used to describe the last days of Adaminaby:

> To drive for the first time along the dead suburban streets is a nightmare experience. Mile after mile of ruined houses goes by in a grim repeating pattern of stark remains. Mostly, they are just sets of stilts, with iron or concrete staircases ending in mid-air. Children's cycles stand rusting among the roofless pillars. Weeds and grass grow waist-

> high in neglected gardens. Kerbside pyramids of twisted metal and timber spilling across roadways wait to be carted away.[12]

Bagot Reserve, which housed most of Darwin's Aboriginal population, was almost completely flattened and the former residents, like most other Darwin citizens, made their way to the nearest standing public building—the local primary or secondary school.

On Christmas Day the federal government appointed Major General Alan Stretton to take initial charge of the relief and rescue operation. He arrived on 26 December to determine which of the many plans and suggestions for the immediate and long-term Darwin crises to adopt and implement.

One of the most bitterly debated aspects of the cyclone's aftermath was the evacuation of women and children. The favoured bureaucratic phrases were 'thinning out of people' and 'every one [remaining] is a mouth to feed'. Between 26 and 30 December some 25 000 people, mostly women and children, were evacuated from the city, many against their wishes.[13] 'Some of the other wives of people with us wouldn't go, couldn't go at that time. It was in fact too close to the disaster.'[14] The only point on which most now agree is that a small number in particular need, such as nursing mothers, should have been evacuated; whether men should also have gone, whether such large-scale forced evacuation really was necessary, and whether everyone should have been allowed to return when they wished, are still matters of sorrowful and angry debate.

The airlift of women and children, though not officially compulsory, was believed by most evacuees to be so. Opposition to removal emerged so quickly that by 28 December Stretton found that he had only 800 volunteers to fill the 8000 aircraft evacuation seats which he had ordered. He persuaded the Commonwealth government to offer free return flights to all evacuees, and the seats were filled.[15] Only as the evacuees travelled by bus to the airport did the full extent of the disaster become apparent:

> All the men outside [the bus] were standing there, and just helping with those getting on the bus. Then there was the most pitiful sound which I will always remember—the sound as we drove off as every woman on that bus started to cry, I included. ... And then just utter silence as we drove through the streets and looked out the window to the ruined city.[16]

Dawn Lawrie, the Legislative Assembly member for the northern suburbs seat of Nightcliff, recalls mothers asking her ' "We can come back, can't we?" "Of course." But events proved me wrong.'[17]

At the airport there were hours of waiting in the rain without toilets or amenities, holding babies and pacifying small children. There was chaos at southern airports as refugees arrived, sometimes after midnight, to be bundled off to hostels. More than 1000 evacuees arrived in Sydney within ten minutes. All suffered, then or later, a feeling of loss without comprehension:

> That time was really terrible ... Love and concern, yes. But no understanding or comprehension, yes, that's the word. As much as you talked, people didn't really understand. They weren't there, and the loneliness of that week, being without my husband, and without anyone who had been there, to talk to, or to be part of was really terrible ... I have never been so lonely, and jealous in a way, of the companionship which those left behind had, even though it was terrible being there. And getting over it together. All of us here felt the same, just that being so alone, with no understanding, no comprehension from the people down here, however kind, and however lovely they were to us.[18]

The evacuation and the decision in early January 1975 to allow only a small proportion to return, remains to Lawrie a lesson in 'how stupid, arrogant and how completely out of touch decision making can become'. For a few weeks Darwin was virtually a city without children, and she 'realized then just what a dreadful act the Pied Piper had perpetrated on Hamelin'.[19] Twenty years later, Thelma Crossby recalled waiting for news about rebuilding while she waited in Brisbane. 'Everything was different, Brisbane seemed like another country. It was only our home town we had left, but it seemed as if Darwin didn't exist and we were foreigners.' After four weeks she was desperate to return.[20] Cramped accommodation with distant and sometimes incompatible relatives, constantly changing addresses, doubt about who would pay for dental, optical, physical and emotional care, and loss of contact with other evacuees made Darwin residents ever more anxious to return. What had happened to property and possessions? Did mortgage payments have to be met? How long would they be in the hostel? Was it worthwhile putting children in southern schools? When could they return? When could they return?[21]

Kass and Peter Hancock remained during the first week to help establish a regional survival centre which became known as the Wagaman Hilton. With the help of citizen military force soldiers and a regular army sergeant they established headquarters at Wagaman primary school. They ran a pipe from the Moil tank reservoir for drinking water, organised holes to be dug for latrines, commandeered a front-end loader to bury rotting material and freezer units from the supermarket, took charge of the tinned supplies, and within a few days were cooking hundreds of meals a day. While the roads remained blocked the Wagaman community, united by common interests begun before the cyclone, kept morale high with a sense of purpose and solidarity. Kass and others offered counselling to shaken truckdrivers who were arriving to announce how many bodies they had buried in mass graves, the largest at a school in a neighbouring suburb. They showed the wallets and necklaces they had taken from the bodies to identify those whom they assumed would later be exhumed and buried decently. On the basis of the different reports, the Hancocks calculated that 250–300 bodies must have been hurriedly buried, never taken to the temporary morgues, never recorded in the official list of the dead.

Setting up similar self-help camps, residents of other suburbs felt a self-confidence and independence against which bureaucratic attempts to rebuild Darwin later were measured. But such regional solidarity, however locally valued, could not long survive. Within ten days a relief column of military-style authority arrived from central headquarters at the far end of the city to strip the Wagaman Hilton of its brief autonomy. Leadership by known and familiar individuals at known and familiar sites was shattered by outsiders imposing the uniformity of elsewhere. An agent of occupation had invaded local familiarities. Uniformed men began killing supposedly wild or deranged family pets. 'In the stillness of our ruined city the sudden sound had the haunting echo of desolation and death.'[22]

At the end of the first week Kass Hancock and her children were evacuated. A soldier pointed a gun at her in jest and said 'you know I'll use it'. Three weeks later Peter Hancock left the Wagaman Hilton to join his family in Adelaide, physically exhausted.

Returning became progressively more difficult. The Interim Darwin Reconstruction Commission granted permission only to males who were in the city before the cyclone, previously employed females now classed as key personnel, and a few others on compassionate grounds provided that they were 'appropriately housed'.[23] Until June 1975 it was particularly difficult to obtain both a permit and the promised free return air ticket. Few mothers would risk the long drive without a guaranteed right of entry at the roadblock. The evacuees waited, while some southern sympathy turned to uninterest or impatience.

Every one of the score of psychological assessments carried out on the victims of Cyclone Tracy concluded that those who were evacuated from the city and did not return suffered the greatest psychological harm. This post-traumatic phase is said to manifest as anxiety, fatigue, depression, anger, aggression, guilt and prolonged bereavement. Such states were often investigated by doctors and psychologists but seldom treated. As after the fires which devastated Tasmania in 1967, the immediate disaster control was handled much better than the psychological disorders.[24] The view prevailed among many professionals that follow-up counselling would only 'emphasize the problems and initiate or perpetuate psychological distress'.[25] Psychologist Gordon Milne found that 31% of evacuees who never returned to Darwin suffered emotional disorders, while the rate of both those who never left and those who returned quickly was below 13%. Evacuees who remained in the south spoke as if they were 'mourning over loss which went deeper than deprivation of house, possessions or job. [Their trauma] expressed alienation from a physical and social environment which is probably unique in Australia.'[26] Twenty-five percent of the evacuees surveyed in 1981 by the Department of Social Security regretted having been evacuated.[27]

The professional carers, as so often in Australian history, totally misjudged the emotional needs of the victims of Cyclone Tracy, but the psychologists got it half right. They identified the terrifying events of Christmas Eve, the disintegration of family and community, the death of friends and post-evacuation experience as causes of psychological disturbance; while the more technical noted the effects of 'transient situational disturbance' and even 'reduced environmental stimulation', very few identified the loss and continued deprivation of a place of attachment as a significant cause of emotional distress. The Department of Social Security survey conceded that 'since homes are of central importance ... every consideration should be given to restoration where possible, or to rebuilding on the same sites to maintain established communities'[28]—without admitting that a contributing factor in emotional trauma had been the initial destruction of that site. A more perceptive psychologist concluded that 'the continual freighting out of disturbed and upset people many days after the event instead of providing local support and treatment first is questionable'. Why, she asked, did no one draw an obvious lesson from the traumatised evacuees from Europe after the Second World War?[29] Why were centres like the Wagaman Hilton so abruptly terminated? Kass Hancock recalled her strongest emotion in Adelaide: *the desperate need to go home*. Only the Bagot and Kulaluk Aborigines seem temporarily to have benefited. They had few material possessions. They belonged to the Darwin region as did few others; most of the land precious to them was already inaccessible because of buildings and fences. Building materials were now lying about and food was distributed freely at schools. The barracks which the army had built on the most sacred site of the Larrakia people was in ruins. Since the authorities expected so few non-Aboriginal residents to return, the time should be right to press land claims upon the government.[30] No one, in Harry Singh's memory, asked to be evacuated from Delissaville. Nobody was. No one carried out any psychological testing on the Aborigines.

When, eventually, evacuees began to return in larger numbers, they found a city different from the one they had left. Long-time resident William Walsh recalled:

> The missed experience, the missed business being away from town at that time, not knowing what went on, not having a part to play in it and, in fact, losing control over your own life in your own town by being absent while other people make decisions and do things. When you come back it's no longer your town, it's been alienated, you've been alienated from it.[31]

The feelings of alienation were the more unnecessary because almost everybody, like Anne Boyd as she rebuilt her house at Ferntree Gully, wanted to be close to where they had lived previously. Indeed, almost everyone insisted on remaining on their block or under the raised floor of their former house, cooking on barbecues or eating in the nearest school canteen. Communities began to re-establish themselves. George Brown, the government horticulturist, recalled driving round Darwin streets issuing two plants to every home. *And people would just come out from under houses ... and take a couple of trees away with them You'd see them hugging the bloody things, you know, and the kids'd come up and say 'Can I have one too?'*[32] Peter and Kass Hancock returned at Easter in a converted bus which they parked on their block. Peter felt annoyed that he had not been able to salvage the building material since cleared by the navy, Kass that she had not had an opportunity to search their block for lost items and keepsakes. They felt that they were in a zone of foreign occupation as navy trucks rumbled by on apparently important missions and men in uniform were everywhere. *It was a force from somewhere else that made me uncomfortable.* The place was familiar if many of the people were not. Kass felt an affection *immensely deep* for the school, the shop site, the whole suburb and its citizens.

For the first month city morale was high. Postage was free, as were meals, pet food, dental, medical and optical services. Water was restored within days, electricity within weeks. A daily government newsletter announced which house sites would next be cleared by sailors. Five-day trading resumed after three weeks, normal air services after a month. By the end of January 1000 houses had been reroofed. Sardonic signs like 'Heartbreak Hotel No Vacancies' appeared on ruined buildings. Observers extolled the therapy of rebuilding. The victims were 'far from recourseless' and to treat them 'as manipulable objects was to impede rather than encourage their recovery.'[33] In the first hours officials had wondered whether Darwin should be rebuilt at all, and even deputy prime minister Cairns wondered whether the city should be reduced to an air force base.[34] Residents who had a greater personal stake in and understanding of their living area were more confident. *Most of the people I knew were just pooh-poohing it and 'Of course it will. We've come back here. We've got to live somewhere.'*[35]

The residents were setting themselves against the phenomenon we noted first in eighteenth-century Britain: each locality was part of a larger area. What was special about any particular place was forgotten or uninteresting; the map was drawn blank by the surveyor. The planners had focused on city-wide projects intended to match Darwin to the cities they knew. Darwin was to become a southern city. A joint statement by the Ministers for the Northern Territory and for Regional and Urban Development early in January 1975 announced the formation of the Darwin Reconstruction Commission to 'set up, develop and reconstruct Darwin', ominously comparing the city to Canberra, Albury/Wodonga and other areas selected by the Whitlam government for 'regional growth'.[36] By February 1975 a second attempt to destroy local and familiar Darwin had begun, only a month after the first.

Darwin was a Commonwealth-funded public service town, more at the mercy of faraway planners than other capital cities. It was not the first time that residents had resisted: Japanese air raids during the Second World War destroyed two-thirds of the city, and after the war the Commonwealth had formed elaborate restructure plans. A city planner then reported testily that the residents had 'objected strenuously' to the proposed changes, even refusing to allow a shop to be moved 100 m from its position.[37] Although waning enthusiasm and departmental disputes brought the plans to nothing, neither housing allotments were released nor building loans approved until the early 1950s.[38]

Discussion was already occurring before the cyclone on the future shape of the city. There were some obvious problems, the most pressing of which was the airport separating the business district from the rest of the city. Now the Darwin Reconstruction Commission had been granted enormous powers. Except for statutory authorities, the Commission was responsible for all construction within 40 km of the city, town planning, civil defence, essential services and capital works, all government buildings including housing, and the standard of private buildings.[39] The Commission seemed an idealistic—though remote—attempt to combine modern city planning principles with higher standards of building codes. To local critics it meant that the Commission had authority to override any existing town planning ordinance, change land tenure titles and veto the erection of everything from individual buildings to entire suburbs. Almost immediately rumours flew about. Was it true that all the northern suburbs were to be bulldozed or were the authorities merely testing community reaction? Only three weeks after the cyclone the Department of Housing and Construction circulated an extraordinary document in which the population was projected at 25 000 until the end of 1976, but a 'problem' was anticipated by people remaining in Darwin wanting 'to share in the determination of the goals and the shape of the re-emergent community'.[40] In passing, the report objected to the 'hundreds of similar looking ... house types advancing in regimented rows in a continuing monotonous sprawl'. The accompanying map blanked out the whole suburb of Coconut Grove and some of Fannie Bay, which were to become 'parklands recreation open space'.

A week later the Cities Commission, a statutory body established by the Whitlam government, produced *Planning Options for Future Darwin*, suggesting a complex for cultural, administrative and recreational purposes, to be designed through a national competition. Only essential buildings would be allowed to be repaired in the historic area of first European settlement. This second plan set the population at 40 000, still less than the pre-cyclone total. It recommended that a fund be established to allow the government to buy back—the planners meant 'forcibly requisition'—housing allotments in areas where land-use zoning was to be changed. It mooted large-scale acquisition of properties and homes, the establishment of satellite cities of North and East Darwin, moderate change to the residential suburbs of Nightcliff, Rapid Creek, Coconut Grove, Ludmilla and Winnellie. Employment would be held to 10 000 in the central business district where major rebuilding should be concentrated in 1975; there should be no complete rebuilding in the northern suburbs at all. Residents from those areas wishing to return should be rehoused elsewhere for up to three years. The planners could 'probably envisage the retention of some existing individual leases provided they could be fitted into the detailed scheme of further development and reconstruction in that area'.[41] The remoteness of the planners from local concerns was reflected in a third government report of April 1975, redolent of the self-confident urban themes of the 1970s, which criticised the city for having allowed residents to live on large blocks. Such disposition produced 'few social gains'. According to this thinking there was little social contact in Darwin, no local experience. 'Within the family dwelling can be found every device to substitute for communal activity: the deep freeze instead of daily shopping, record players instead of concerts, tv sets and telephones instead of face to face contacts.'[42]

Now Darwin families all over Australia were clamouring to be allowed to return and some had succeeded. By mid February the population had doubled, and nearly 3000 blocks, some 35% of the total, had been reoccupied. The new arrivals listened in alarm as Dr Patterson, Minister for the Department of Northern Territory, proclaimed that the new Darwin might not be the 'Darwin of old', for 'we' had the 'opportunity of creating a city'.[43] Twelve hundred objections were lodged to the Darwin Reconstruction Commission's plan. A second report was issued in March in which a more circumspect Minister for Cities recommended a more cautious approach. Minister for Northern Australia Uren noted the citizens' rejection of some of the recommendations, especially for medium-density housing. Instead they had showed an 'overwhelming desire ... for rehousing on the same block, preferring to remain under tarpaulins there than in a caravan elsewhere'. The planners now conceded that it would be impossible to carry out the reconstruction of the northern suburbs because of the surviving infrastructure investment, intact buildings, legal difficulties and reoccupation by 1000 residents.[44] Plans remained, however, to drive new roads through old suburbs, to close streets and to relocate shopping centres, a community college and a school.

The question of whether the northern suburbs should be rebuilt turned on the 'surge zone', which was the area of land expected to be inundated by the sea if the next cyclone struck at high tide. After division into primary, secondary and tertiary zones, the blue danger area on planners' maps extended over 900 housing sites, almost half the northern suburbs. If it was so dangerous, residents asked, why had they been allowed to build there before the cyclone?

By March the suburbs were reconsolidating. Barbara James remembers a Nightcliff community planning meeting dissolving in uproar when a Reconstruction Commission official proposed demolishing the only remaining house in a street in order to provide a public park.[45] Like many others, Peter and Kass Hancock ignored authority and rebuilt their Wagaman house without plans. A sense of camaraderie and common purpose was sharpened by opposition to the common enemy, the Reconstruction Commission. It was strengthened, in Kass Hancock's case, through the *immense freedom and a sense of total invincibility* which flowed from having survived the night of appalling terror. She loved the *intense involvement* in returning to community concerns, the power—or the illusion of power—of having a say in the decisions of local democratic life. *I felt like I belonged to the city and the cyclone made that feeling.* Peter Hancock recalled *We wanted to retain our autonomy in the suburb, we thought we'd keep that feeling. This was going to be the new world.*

Throughout 1975 planning proposals, some well-intentioned, were interfering with local autonomy. Buffer zones one allotment deep were set aside on each side of major thoroughfares, as plantations intended to catch housing debris during another cyclone. Owners of destroyed properties in these zones were allowed neither to rebuild nor sell their land. Gerry Tschirner, who lived in Trower Road, twice wrote to the Commission to ask if he could have another block. Neither reply nor acknowledgment was received.[46] Tschirner commented in 1994 that 'Darwin, despite its appearance, was not a blank sheet of paper on which the planners could begin afresh.'

Opposition to Canberra plans increased. Denys Green wrote to protest at another 'politically oriented, grandiose, socialist-based rebuilding organisation foisted upon Darwin. This is a people's place not a planner's paradise.'[47] Anglican minister Keith Cole considered that the 'huge monolithic monument' of the Commission was 'bureaucracy at its worst'.[48] Grant Tambling MP told the Legislative Assembly, 'This is an existing community with an existing social framework; it is not a growth centre, a new town or an expansion program or an experiment to be played around with.'[49] Most outspoken of all was the member for Nightcliff, Dawn Lawrie, 80% of whose electorate was to be evicted on the first proposal that the suburb be turned into parkland. She described the Commission as 'an octopus reaching its tentacles into every fabric of our life ... a leech living on the life blood of the community'.[50] Even the moderate Harry Giese, the former Director of Welfare, asked why such entities as the Entertainment and Sports Committee were established when such committees already existed.

By the middle of June 1975 there were still no firm guidelines for reconstructing damaged houses. No loans were available and penalties applied to those who, from their own resources, exceeded the specified financial limit. As criticism of the Commission mounted, its chair, the builder Sir Leslie Thiess, resigned and was replaced by Tony Powell, the head of the National Capital Development Commission in Canberra. Almost all the members of the Commission lived in Darwin, Canberra or Melbourne. The domination of local residents by outsiders reached its peak when the elected Northern Territory Legislative

Assembly voted to abolish the permit system, on 27 March, but was overruled by the unelected Department of the Northern Territory. The Assembly, it argued, had 'voted to endanger the lives and health of tens of thousands of people'.[51] The permits remained in force until June, by which time many people had begun their lives elsewhere. 'Perhaps only one quarter of the original population of Darwin has returned.'[52]

Dawn Lawrie founded a Residents Action Group in her Nightcliff electorate, at first to encourage self-help, but soon to fight the plans to eliminate or rezone the suburb.[53] By May 1975 there were twenty similar groups in Darwin and several others in southern cities sponsored by residents still unable to return. Some simply called themselves protest groups. And community opposition, aided by the mounting costs of reconstruction, succeeded. By the end of February 1975 the first plan to obliterate the northern suburbs had itself been obliterated. By June, repairs to existing buildings were allowed to proceed; by August houses could be rebuilt on their original site and construction on one side of major roads was permitted. In August 1975 Peter and Kass Hancock, following the example of the Resident Action groups, founded a community newspaper called *The New Darwin* by and for 'the community at the ordinary level of advising, helping, chatting, getting angry, laughing, and occasionally crying'. The lead article of the first issue called all Darwin citizens to arms: 'Instead of allowing doomsday prophets and knockers to demolish us, it's about time we struck a blow and that means as an entire community.' In later numbers medium-density housing was emphatically rejected, and opinions were expressed that even former residents who had not yet returned should have a voice in planning.[54] Capitalising on the Whitlam government's sundry crises, the community groups continued to subvert development schemes. By September 1975, 900 house sites were reallocated in the surge zone; by December block owners were told they could build where they liked so long as the floor of the house was above the primary surge zone; the main road buffer zone plan was abandoned. In December 1976 the Darwin Disaster Welfare Council, on which sat a score of the city's senior residents, passed a preliminary verdict on the Darwin Reconstruction Commission. They judged that the Commissioners, 'heady with the thought that they could plan the city from the rubble', had 'completely overlooked the emotional investment of the people in their own block of land' and 'displayed no understanding or even sympathy for the basic processes of participation which one would correctly assume when rebuilding not on an empty sheep paddock, but over the ruins and half ruins of family homes'.[55] They conceded that a certain amount of replanning had been desirable and would have had general community acceptance, but as a result of the planning overkill, replanning had been rejected out of hand.[56] Lawrie commented 'The only thing they had left on Christmas Day was the land ... our fate was being decided by people who'd never lived in the place.'[57] Though granted a five-year term, the Commission disbanded itself at the end of 1977.

The human cost remained. Suzanne Parry was chagrined, first because her family's house site in Coconut Grove had been requisitioned by the government, second because the block had later been sold to someone else; and third because house building was now permitted for another 200 m past their block almost into the mangroves at the sea's edge. Gerry Tschirner was eventually allowed to rebuild along the edge of Trower Road—after he had been separated from his family for a year. By the end of 1975 the almost ecstatic sense of self-help began to evaporate as homes were completed and people resumed their normal jobs. The volunteer work-gangs were unfilled:

People started to get their places rebuilt and all of that camaraderie, all of that friendliness started to disappear. The things that started to separate us were the fences started to be put up again and people started to get their lawns down and mow them. So we went back to suburbia as it used to be where people wouldn't talk to each other very much.[58]

The local associations failed to fulfil the highest hopes. Yet while the city socially and culturally remained disorganised and drastic, Darwin itself survived as a much tighter unit than other capital cities. The eroding self-confidence of the suburbs dissolved into the wider solidarity of a whole city fighting the instrumentalities of the outsider. Those who had had the perseverance and good fortune to remain or soon return were the stronger for it. A psychologist found that the suburban residents who regrouped in Darwin formed less complex and culturally defined primary groups, more informal, more intimate and much more therapeutic than the former residents who remained in the southern cities, scattered, lonely, uncounselled and uninformed. Those with the highest psychosomatic disorders, emotional disturbance and addiction were those who did not or could not return. It was best to remain in the 'impacted community and be subject to its integrative and regenerative forces'.[59]

Twenty years after the cyclone it was clear that the city itself had been victorious over the intentions of an outside and totalising authority. In this book of lost places Darwin remains the only example of insider success, not by a home-owner, suburban action group or threatened town, but by a whole city.

REMEMBERING THE CYCLONE

The traces of Cyclone Tracy are still plainly visible in Darwin. At the seafront at Coconut Grove and Nightcliff, and even in the central business district there are vacant blocks, weedy driveways, faded shreds of canvas awnings and overgrown fences. Almost all carry a developer's hoarding: Darwin is no Macedon where the sad and forgotten remains of environmental disaster lie hidden in the long grass. In time Darwin citizens will rescue all these empty blocks from the past. A few 'flat tops' remain in the suburbs, the exterior stairs leading nowhere. Once they were the floor of raised houses, then aching ruins; now they serve a new purpose of ground level garage or workshop.

The people officially listed as having perished in the cyclone are remembered in several locations. The names of seven trawlermen are listed on a somewhat obscure granite block at Christ Church Anglican Cathedral, and those of all known victims are inscribed at the Darwin Civic Centre. Several schools have simple memorials to children who perished. One reads:

WAGAMAN PRIMARY SCHOOL
1/12/76
In memory of the Children
who lost their lives during
Cyclone Tracy 24/12/74
Opened 1973
Tracy 1974
Reopened 1976

Compared to the countless stories of devastation and escapes told and retold, little is known or discussed about such deaths. Not much was reported at the time, and Major General Stretton, interviewed by the *Australian* on the cyclone's twentieth anniversary, revealed a number of gruesome details which he had perhaps forgotten that only the higher authorities knew.[60] The relatives of victims kept, and still keep, silent while almost everyone else who endured the hurricane has told the story again and again. A woman who consented to be interviewed during the commemorations was more distressed by later events than she was by the cyclone itself. Sherylee Armstrong related that her mother had been killed by a flying splinter, but the first she knew of her mother's name being placed on the Civic Centre list of victims was when she received a photo of the Queen opening the memorial. There had been no phone call, no counselling, no information: the authorities had not even bothered to find out her mother's middle name.[61]

The most common physical commemorations of the cyclone in Darwin are several ruins left to demonstrate its ferocity. The desire of the Reconstruction Commission to clear away the wrecked original buildings of the old town was countered by the residents' strong desire to keep the little that had survived.[62] The stone entrance of Christ Church Cathedral, the only part to survive, has been retained and built into the glass and steel structure of the new church. The partly demolished town hall has been left in spectacular ruin, its floor rebricked in whirling concentric circles as if still held in the eye of the hurricane. The concrete floor of the laundry block of Fannie Bay jail is retained as a reminder of the cyclone, and one or two other totally destroyed sites, like the Retta Dixon Aboriginal Children's Home, are commemorated by plaques. A strange steel entanglement outside the Casuarina High School bears the ambiguous inscription

Three House Girders
Twisted During
Cyclone Tracy
25th December 1974[63]

Much controversy surrounded the twentieth anniversary commemorations in 1994. Among the less contentious productions were a book *Cyclone Tracy: Picking up the Pieces* by oral historian Bill Bunbury,[64] a Christmas card produced by the Northern Territory University illustrating its damaged and rebuilt site, and a strictly factual series of information sheets published by the Darwin Public Library and Information Service. Less tasteful was a T-shirt produced by Australia Post which read 'I survived Cyclone Tracy' featuring a caricatured male sitting among rubble with a brick on his head.

The Anniversary Committee established to create and co-ordinate activities attracted controversy from the first. Was the commemoration to be a celebration or a mourning? Symphony concerts were acceptable, but some survivors questioned the appropriateness of a cocktail-mixing competition and a 'Rock Down the Wind' concert. The committee chair claimed the occasions were to celebrate 'how well Darwinians have rebuilt the city', and compared the plans to the delayed homecoming of Vietnam soldiers. 'It took me a while to come to terms with and accept ... but there is a time to let it all out of the way and that's what I think we are doing.'[65] A *Northern Territory News* journalist requested survivors' accounts for a commemorative supplement: 'I see no conflict between dignified commemoration of the event and joyful celebration of what has been achieved since.' Sherylee Armstrong, her memory revived by the interview, told the newspaper:

> I think it's cruel, really I do. It digs up a lot of feelings a lot of people keep hidden. Twenty years down the track your lives have changed, you have families and the last thing you want is a tragedy thrown in your face. I don't know anywhere in the world where they celebrate a bloody tragedy.[66]

A second correspondent commented 'I guess the East Germans were too dull to see the potential in celebrating the fire bombing of Dresden.'[67]

On 30 November, the day before it published a forty-page 'cyclone special', the *News* printed a lengthy explanation of its policy of remembering the destruction of Darwin: the staff had discussed the issue among themselves; they were particularly conscious of its sensitivity; the supplement would awaken many memories people would prefer to remain dormant. The newspaper respected these feelings, but 'the personal accounts, untold before, must be published'. Above all, few of the survivors had had much faith that the city would be rebuilt at all. 'In large part it is the achievement of the people who survived and put their faith in themselves and their neighbours.'[68] Three-quarters of the supplement was devoted to survivors' accounts of the fateful night; short articles on looting, the evacuations, problems of non-returnees and rebuilding the city made up the rest.[69] The phrases 'Triumph of the People's Spirit' and 'The Story of the Ordinary People of Darwin' separated the experience of the night of terror and the evacuation from the other intense and enduring traumas—death or absence of friends, relatives and neighbours, prevention of return, the plans of the Reconstruction Commission, feelings of living in an occupied war zone—and the almost total destruction of the cultural and social life of the city. Kass Hancock, in faraway Canberra, heard of these deep divisions and pondered their meaning:

I think it reminded me of the early days after evacuation and I suspect there was sense that the local people ... [felt] that again that it was somebody else doing it, it was being done for them and it wasn't accepted as such.

The sensitivity of public memory was illustrated in a fiery exchange between two survivors. Kevin Rainbow of Queensland, who had lost shipmates in Darwin harbour, wrote to complain of the celebrations: 'to think that a group of money-hungry individuals can attempt to make money out of these people's loss and grief, makes my stomach turn'. A Darwin resident, Pam Flint, replied:

> I don't think the hard-working committee will be retiring to Spain ... on the profits. Or are you just put out that you didn't think of it first! ... I certainly celebrated being alive on Christmas morning 1974. Were you here for it Kevin? Did you have to rebuild your home, your business and your family's lives as a result of this horrible nightmare? Have you worked your butt off to get where you are today? I think not ... So Kevin stay in Queensland. Leave Darwin to the Territorians.[70]

Also controversial was an exhibition mounted by the Northern Territory Museum. Its co-ordinator, Dr Mickey Dewar, intended that the presentation not stress sentiment or emotion. She wanted visitors to understand the different phases of Darwin history, to make connections between the sites of 1974 and 1994, and to take cyclones seriously. Even the decision to begin was controversial: half the people who contacted her advised *Don't do it: leave the past where it was.* Dewar replied: *These events will always be remembered but it is how we remember them that gives meaning to the present.*[71]

Visitors to the exhibition were presented first with the artistic responses of two Aboriginal men, Geoffrey Mangalamarra and Rover Thomas. Almost everybody, Dewar recounted, had initially considered an Aboriginal response as somehow extraneous to the disaster. Like the display of photographs of Macedon, visitors then inspected a series of triple views, before the disaster/just after the disaster/today. They proceeded to a wreckage-strewn garden. Pieces of masonite, broken timber and the end of a fallen Christmas tree seen from the bottom of the staircase invited the imagination to fill out what might lie on the hidden 'flat top'. The technological centrepiece of the exhibition was the continuous playing, in a lightless chamber, of a recording of the cyclone made towards the end of midnight mass at St John's Church on Christmas Eve. The sound of rain and high-pitched whistling were added to the original frightening deep rumble, and visitors were invited to lean against a piece of corrugated iron which vibrated at moments of highest aural intensity. Former Legislative Assembly member Dawn Lawrie was the fiercest critic of the exhibition, as she was of everything associated with the commemorations. She thought the entire program macabre, totally inappropriate, repulsive, repugnant. For visitors, exhibitors, journalists and historians she had only one message: *Will you please leave it alone.*[72]

Peter and Kass Hancock, who had left Darwin in 1978, returned for the commemorations. On the morning of 24 December 1994 Kass Hancock learnt that the authorities had at last conceded that in addition to the sixty-four named dead, 245 people were officially missing. The figure tallied almost exactly with their own count of hurried pit-burials in the first days after the cyclone. *It was one of the most important moments in my life.* Even if their names were never listed nor a memorial service held for them, *at least they existed.* Wagaman itself was almost unrecognisable. The rebuilt houses were of a different style; trees concealed and transmuted what few buildings remained. Casuarina High School seemed much smaller. None of their immediate neighbours remained, only two elderly ladies from the other end of the street were familiar. The preschool, now the child-care centre, was barely recognisable. The Hancocks tried to find their way about the supermarket and were soon disorientated. *It was the same but it wasn't, like something which reoccurs in a dream to you, partly familiar, but it's not really, and it changes.* The preschool was virtually the only physical reminder of the suburb which had vanished as comprehensively as Adaminaby and Yallourn, and which, for a few months in 1975, *was going to be the new world.* Kass Hancock thought of a house across the road where two little children had been killed by falling concrete. *I looked and I couldn't even pick what block the house had been on because that area was so different.*

Of old Wagaman, only ghosts remained: *I'd been back a lot to that school in the time after the cyclone, and there was always ghosts, there were the faces I could see of kids that I had in my pottery class who'd been killed, mainly in playgroups, kids that I'd handled and had in my house.* December 1994 was a time to lay them. Peter Hancock's ghosts were personal and relational, Kass's ghosts were for deaths unacknowledged: *knowing that you have to go back and grieve because there wasn't time.* The Hancocks attended the memorial service on Christmas Eve. To Kass Hancock the service *was something like a burial, a rite of passage. Something that should have been done but couldn't, had been done.* She found that feelings for the suburb were:

immensely deep for me. The whole suburb, the school, where the local shop had been, even a couple of cracks in the footpath, given that all the built environment was gone, a sort of tremendous association with the place was walking over that footpath and remembering those cracks that were there when the kids started school I walked down here.

So the only artefact remaining in the whole of the rebuilt city which Kass Hancock could recognise, kneel down and touch with close familiarity was a broken pavement in Wagaman Terrace. Secure against the hurricane, secure against developers, the cracks were smaller than a town, a house, a garden, smaller even than a plant. On them or in them she felt a surge of recognition, a physical reminder of 1970s culture, relationships, family, feelings, work, all that had been destroyed—there on the cracked pavement, that place.

Perhaps aided a little by the disarray caused by the cyclone, the Larrakia Aborigines won the Kullaluk land claim.[73] The Aboriginal community of Delissaville still lives in the temporary structures built a few months after the cyclone, as the promised permanent housing has never eventuated. In April 1995 the Aboriginal journal *Land Rights News* photographed Harry Singh beside the ruins of the 'hot boxes', the crude tin pre-cyclone shelters erected by the old Native Affairs Branch and never removed.[74]

The non-Aboriginal citizens preserved all their suburbs and most of the city centre. Their victory was aided by political confusion, the expense of comprehensive rebuilding and the fact that the Reconstruction Commission's plans were driven by an expendable ideal rather than, as in the case of Yallourn, an apparent economic necessity.

In chapter 8 wc see another fight in urban Australia as the residents of Beecroft, New South Wales, tried to retain the integrity of their local culture and place against a monolithic opponent: and lost.

CHAPTER 8 LOSING A NEIGHBOURHOOD

Lost neighbourhoods are lost people:

> I don't see too many of my friends in this neighbourhood. I see some of them, but there is no togetherness here as it was. We felt safer, we felt like we were in a Greek town. You came out from your house into the street and you met so many people who would say hello to you ... [After slum clearance in Chicago] so many things changed for the better. But for individuals—now all these people who lived down there, I don't think any of them is happy now.[1]

Lost neighbourhoods are lost suburbs:

> The Chicago I knew was vast and squalid ... Slum clearance hasn't improved it ... It is not a shanty town any more, but possibly something worse ... While no one regrets the vanishing of the old slums, we also remember we once had neighbourhoods. They have vanished too. Without them, there can be no such a thing as a city to which one feels held. We are passing into a city without roots. We must conquer the city or be conquered by it.[2]

Lost neighbourhoods are lost artefacts:

> There was a Japanese elm in the courtyard ... It used to blossom in the springtime. They were destroying that tree, the wrecking crew. We saw it together. She asked the man whether it could be saved. No, he had a job to do and was doing it. I screamed and cried out. The old janitor, Joe,

> was standing out there crying to himself ... At night the sparrows used to roost in those trees and it was something to hear the singing of those sparrows. All that was soft and beautiful was destroyed. You saw no meaning in anything any more. There's a college campus on the site there now.[3]

Almost all the Australian and overseas critics who have attacked suburban life have missed the point of these laments. The English cultural critic James Froude in the late nineteenth century believed that a purely urban (including suburban) life threatened a people with degeneracy. The American architect Louis Esson decreed that suburbs were devoid of joy, ecstasy and spiritual adventure. Marxists held that suburban culture was 'inauthentic' because it fostered a false sense of individual freedom.[4] Dozens of critics, oddly eager to criticise those same areas where often enough they themselves lived, damned Australian suburbs as 'beastly little bungalows'[5] 'crowded with flatulent, faceless men'.[6] Chris Wallace-Crabbe wrote in 1963 of Melbourne's 'hundreds of square miles of suburbia into which the population flees in the evening, draws down the puritan blinds and settles itself before the blue shimmer of the television set.'[7] Writers A.D. Hope, Barry Humphries, Dorothy Hewett, Robert Drewe and Jack Davis are among those who have slated suburban living for its apparently self-evident meanness of spirit, sameness and anti-intellectuality.[8]

The artists who shaped much of Australia's cultural position adopted suburban alienation as a cultural theme; if the bush was weird then the suburbs were meaningless. Cultural historian Bernard Smith searched for but discovered very few artists before 1945 who portrayed suburbia as a meaningful part of their personal environment. A few writers and painters, including Sali Herman and David Malouf, have seen value, identity, warmth or affection in the suburbs; but many of the critics, in historian Alan Gilbert's words, confused and conflated domesticity and philistinism.[9] Just as Dickens' powerful conceptions of the anonymous dark city became stale and conventionalised towards the end of the nineteenth century, Australian writers fell too easily and uncritically into the cliche that Australian suburbs were boring, anonymous or mindless.

The urban planners who influenced the shape of early twentieth-century cities did not intend them to be so. The American landscape architect Olmstead, for example, believed that the middle-class suburbs he planned were a delicate synthesis of town and wilderness.[10] Another American reported that Australian cities 'spread out into the suburbs in a splendid way. For miles about are broad roads, with small houses, gardens, and an opportunity for touch with the freer, sweeter life which the country offers.'[11] Urban Australians who could afford it rushed to the outer suburbs from about 1910; in the first two decades of this century, every state government established schemes to allow all married Australians their own house and land. By 1920 all states had passed legislation allowing banks to offer long-term low-interest housing loans. State-of-the-art suburb design followed the 'spiderweb' principle (a radius round a centre rather than a geometric chequerboard). Rosebery and Haberfield were two Sydney suburbs designed following the latest American and British styles, with the home-grown addition that they were intended to offer a fair go for families regardless of income. Model working-class suburbs like Daceyville were intended to produce good quality housing, available for rent if necessary, with a playground for every twenty houses.[12] These projects too missed the point emerging throughout this book: that people are apt to sink their roots into, and form attachments to, any place they find themselves. Beyond the self-constructed units of the house, the house–garden, the house–garden–street, the house–garden–street– shops—all of which have their own self-perceived unity—lie the neighbourhood, the territory, the suburb, the physical space and the social community.

The enemy of working-class suburbs is slum clearance. The sociologist Herbert Gans studied slum clearances in Boston. The west end, long a home of native-born Americans of Italian descent, was declared a slum and torn down during the Federal Renewal redevelopment schemes in 1958–60. Seven thousand former residents were dispersed over the entire metropolitan area, while the area was rebuilt as high-income luxury apartments. Gans was critical of every aspect of the forced evictions. First, the possibility of destruction had been discussed for so long that it had acquired a false air of inevitability; second, the implication that poor living conditions caused anti-social or pathological behaviour; third, that no distinction was drawn by urban planners between harmful and merely inconvenient features of domestic life; fourth, that middle-class planners assumed that run-down houses affected the residents' perception of themselves as respectable citizens; fifth, that the real reasons for slum clearance were the west end's shrinking retail centre, the hospital's desire for clients with higher incomes, the city's need for a higher rent base and the pressure of developers. No developer or planner heeded the cries of the dying Boston suburb: 'The place where you're born is where you want to die; it pulls the heart out of a guy to lose his friends.'[13] Gans estimated that probably 1 million Americans lost their urban and suburban living areas between 1950 and 1980. Though the era of the public bulldozer has now passed, American governments still find ways to evict and demolish. He noted, in an aside of sinister significance in relation to this chapter, that his calculations of the unhoused did not include the destruction of suburbs by expressways.[14]

Left their houses weeping and became unemployed. Australia has suffered from slum clearance, though less than in the United States where more people live in the inner cities. Serious clearances began in the 1870s, and generally nothing is known about residents' thoughts about their forced removal because nobody asked them. Evidently there was much futile resistance by poor and powerless residents; successful resistance to Sydney slum clearance of the 1880s came from those opposed to the city corporation's coupling of opium-smoking and prostitution with run-down housing, and from the owners of the rented properties.[15] The briefest of glimpses of attachment to inner-city suburbs comes from a 1911 petition signed by 600 residents of Chippendale, who protested that it 'would be very unfortunate to have the city proper largely filled up with factories and the people turned out'. The petition was signed by many people of the same name, living next door to each other or in adjacent streets, 'in what was clearly a "community" '.[16] Chippendale was barely recognisable after the city council's 'beautification' and 'improvement' (read 'slum clearance'), which finished in 1912. At about the same time, resumption in the Sydney suburb of Ultimo demolished a whole suburb—many streets vanished, 435 houses were demolished and 1779 people whom the city health officer described as an 'undesirable class of tenants' were uprooted.[17]

Clearances occurred in the optimistic years of the early twentieth century. This was the other side of the coin of suburban planning. Between 1905 and 1912, 7000 people of inner Sydney were unhoused without compensation. Did the residents care when their houses were demolished? Certainly they did when the corporation, like enclosing English landlords, made no provision for their rehousing. What did they take with them when they went? Where did they go? How attached were they to Ultimo? Julia Martin, born in Germany, lived in one of the houses built to replace 'old' Ultimo, which was itself demolished forty years later:

If I think of my home it is Sydney, not Australia. If I'm feeling miserable I can just go into the city and I feel revived. Even walking about Ultimo and the old powerhouse [now the Museum of Technology], that's really my home, that's where I feel. It's not a national thing. My brother and I used to just disappear and look through everything [in the older Museum of Technology across the road]. We lived in Wattle Street ... We had to move out, and I think it was pulled down and made into a garage. I'm really happy to go back to Ultimo and Glebe. I've worked as a volunteer in the Powerhouse Museum ... I'd be happy to live there. I'd live right on the steps of the Town Hall right in the middle of the city, unhealthy as it is, I would. I was living out in the suburbs, and I hated it.[18]

There are more recent memories of slum clearance in the inner-city suburbs of Melbourne. An older Carlton resident, Amy Phillips, recalled:

> before we were shifted into this Housing Commission flat, I lived across the road in a street called Cross Street, which doesn't exist now, we had a two-storey house, with a garden and an apple-tree and a barbecue area out the back ... We lived there for nineteen years, went to work and the kids went to school from there, my husband worked here, I worked here, the kids went to school here, all their friends were here, Carlton was our home, our suburb.[19]

No pets or gardens were allowed in the new flats for the compulsorily rehoused. Three pubs had gone and were not replaced. Residents disliked waiting for lifts, not being able to do the washing when they felt like it and being unable to leave their front doors open. The new houses cost nine times as much as the old. Why had such drastic urban resumption been necessary? Matters became clearer as construction for Melbourne's F19 Freeway began.[20]

Carlton, Ultimo and Chippendale had changed almost beyond recognition, but few outsiders, though attached to their own special places, wanted to understand the emotions which residents felt towards their dying suburbs. Similar processes occurred in the Melbourne suburb of Fitzroy. On the land now occupied by the Atherton Estate, no streets existed for the residents to return to, and the culture of the surrounding streets became dominated by yuppies and the cafe circuit of Brunswick Street:

> Some may wish to dismiss their [the former residents'] sense of loss as expected nostalgia or sentimentality. But those who lived in Fitzroy had, despite internal hostilities and outside intervention, created viable 'communities' for themselves. When it was taken from them and literally destroyed the loss was real ... *There isn't a scrap that I can point to that I can say 'that's mine'*.[21]

The first suburbs to be rescued from the outsiders' totalising gaze were the working-class inner-city suburbs which had received so much attention from urban planners. Urban attachments were expressed by local authorities as well as by individuals:

> I believe in the Place where I live ... I believe in her people, in her girls, in her boys ... I will spend my money here, and by doing so leave a part of my purchase price to circulate in the channels where its equivalent in wealth was originally created, to do good among the people who are part of the place in which I am a part, in the place I call my home.[22]

Through a series of well-publicised green bans, a combination of unionists and residents of Woolloomooloo and Glebe defeated Sydney slum clearers in the 1970s. In the 1980s working-class suburbs were examined by film-makers and artists as sites of interesting human involvement and creativity. The feature films *Man of Flowers* and *Malcolm*, set in Melbourne, utilised the urban environment as a structural element in the action. The new-found valuation of working-class suburbs by outsiders contrasted oddly with the parallel uninterest in the preservation of middle-class neighbourhoods.

What, then, are these entities we call suburbs, valued by residents if not by outsiders, always under threat of oblivion? Often they segregate themselves by race, work and social status.[23] Their distinguishing characteristics are their peripheral location, low population density, architectural similarity, easy availability, and economic and racial homogeneity.[24] Another distinguishing feature of Australian suburbs is the garden. Suburban planners intended gardens to be a haven beyond industry and the workplace, a civilising and quietening force.[25] Then something else occurred beyond the planners' blueprint. Gardens became not only the sites for 'carpentering, digging, hosing, basking, horsing about'.[26] The gardener's personal association with his or her garden gave it a uniqueness perceptible only to the gardener who made it. The activity bonded the gardener to the garden. Mrs Rolf Boldrewood wrote of her garden in 1893: 'Whatever the soil or climate, nearly everything in my garden has been planted with my hands, and, in consequence, every flower, I may say every leaf or shoot has been specifically known and familiar, and, therefore, more highly valued.'[27]

In the 1930s and 1940s the tendency towards withdrawal from public life became more pronounced in garden architecture. Edna Walling and Jocelyn Brown, in Melbourne and Sydney, influenced design both in their commissioned works and in the journal *The Home*. Their gardens were to be not only admired, but lived and worked in. Brown used phrases like the 'true gardener', 'true connoisseur' and 'the sensitive or cultivated eye'. To her, a garden 'is a place to live in, as well as to practise the art of horticulture; a series of open-air rooms; a place in which we plan, read, walk with our friends or meditate in solitude'.[28] By 1940 the Australian suburban neighbourhood, the community, the streetscape, the home and the garden were established as a site of first attachment, a source of intense pleasure if unhindered, anxiety if threatened and grief if destroyed.

Sociologists who have analysed Australian suburban homes and neighbourhoods as popular culture, style or gender/power relations generally miss the point of these deeply personal associations,[29] and until very recently the critics, not the residents, have held the microphone and the camera. Walter Murdoch hated 'the awful sameness of Melbourne's suburban streets, with their red-tiled houses, neat lawns, gravel paths, *Pittosporum* hedges, eliciting a uniformity of spirit, a complacency, a positive fear of originality or difference.'[30] And the English visitor D.H. Lawrence has been quoted about suburban living far more often than any resident:

> all these little dog kennels—awful piggling suburban place—and sort of lousy. Is this all men can do with a new country? Look at those tin cans! ... Yet there they stood like so many forlorn chickens—houses, each on its own oblong patch of land, with a fence between it and its neighbour. There was something indescribably weary and dreary about it.[31]

'Suburbia' is so familiar to the resident, so unfamiliar to the outsider, so individual to the gardener, so uniform to the critic.

BEECROFT DESTROYED

Beecroft is a bushy, hilly suburb in Sydney's north-west. The western side is on the top of a ridge, divided by Devlins Creek which has cut a gully through the Sydney sandstone and Wianamatta shale. Between the Pennant Hills Golf Club and Murray Farm Road is old orchard land. In the early 1960s agents of Sir Garfield Barwick, the owner of the estate, sold most of it for subdivision, advertising it as 'beautifully located in the secluded natural bushland area of Beecroft ... one of the finest estates in the Sydney Metropolitan Area'.[32] Buyers seeking quietness and a bush environment had architects design graceful but unpretentious two-storey residences. The new neighbourhood became homogeneous in profession, ethnicity and income level. In the 1960s western Beecroft had a local school, parks and shops to walk to; children rode their bikes on the roads and grown-ups had picnics in the bush.

At the beginning of 1994 western Beecroft still retained much of that character. Though the empty blocks had been filled and the roads sealed, no closer-settlement housing had followed the 1960s development. In the gully grew blackbutt, angophora (smooth-bark apple), turpentine and Sydney blue gum; still surviving were brushtail and ringtail possums, flying foxes, echidnas, honeyeaters, thornbills, parrots and snakes, now joined by weeds, carp, black rats, house mice, rabbits and a sewer line. The sugar gliders, wallabies, kangaroos, bandicoots and antechinus had vanished.[33]

Soon there would be much less. The New South Wales Roads and Traffic Authority proposed to build an 11 km freeway from North Ryde to Pennant Hills Road which would destroy dozens of homes in western Beecroft, devastate Devlins Creek and obliterate the neighbourhood. At the time of writing, the compulsory sale of all the homes to the Roads and Traffic Authority is complete and all the former residents evicted; down in Devlins Creek, the bush regeneration enthusiasts have ceased to gather on Sundays. For the first time in this book we consider a deeply loved site under great threat—dying but not yet dead.

Western Beecroft is an area long humanised. There are Aboriginal sites along the creek, stone pathways of the 1820s, buried tracks and roads from the 1840s and 1920s orchard trees in modern back yards. These sites of previous occupation are forgotten or ignored, except by the residents who relax in the bushland, admire the wildlife, draw, write poetry, regenerate the earth, have picnics, or follow the paths to and from Beecroft station.

It was to Mahers Road in Beecroft that Esme Blackmore and her husband John came in 1964. They gave themselves a choice of buying a stone house near the harbour or designing their own house in the bush, and chose the Beecroft site overlooking the golf course to the north and the flatter land of Carlingford and Epping to the south. It was the first and last house they were to own together. John Blackmore, a solicitor, inquired of the government whether the land would ever be required for resumption: the Department of Transport replied that a 'county road' was planned to follow the route of Mahers Road, 40 ft (12 m) wide, in a cutting and close to the golf course. It would be a minor inconvenience; 'no part of the property was required for the Department's road proposals.'[34]

The Blackmores' architect, a family friend, designed a dwelling to make best use of the views, with an internal solarium, open fireplaces, a sewing room, lots of cupboards and bookcases. The floors were tallow wood, secret-nailed with copper. A wooden staircase from ground level to the living areas upstairs was constructed of a cantilevered white softwood with contrasting darker stairs. Outside, John Blackmore, an amateur botanist, began landscaping the lawns, the waterway and the Japanese rockery. The rear garden was designed round huge bluestone blocks brought from the Homebush abattoir then being demolished. After twenty-seven years' residence, Esme Blackwood reflected that she had needed a little more room when the children were growing up, otherwise the design had been perfect. Beecroft was 'green, quiet, and a little haven'. *This is the house for the rest of our lives.*

Ivan and Jenny Lewis lived 200 m further down on the edge of the creek, in Lynbrae Avenue. After occupying fourteen different houses in twenty years, Ivan Lewis moved into his home, designed by a student of Frank Lloyd Wright, in 1965. He built the extensions to the house; the solid garden walls he constructed over many years from stones taken only from the property; he designed and built the surrounding paving. *There's so much of my personal life in here.* The Lewis's back fence abuts the native bush whose restoration Jenny Lewis has co-ordinated. Ivan Lewis loved the bush and its wildlife: over the years he counted nine species of parrot, Jenny Lewis more than 200 native plants. She planted forty red cedars along the creekside. *It's going to be hard to leave that.* Here, for both Ivan and Jenny Lewis, was an *ambience without price ... the red of the angophoras, the rough turpentines, the light shining on the blackbutts—beautiful!*

Ivan Lewis: *If you know where to go, you find huge angophoras and huge blackbutts which would be many hundreds of years old ... with circumference at breast height of 11′ (3 m) so they're more than a metre through. They would certainly have been trees growing at the time of European settlement here. Some are up to 45 m, which are not the typical angophoras that you see growing down on to the coastline, instead of being twisted and gnarled ... these are dead straight. They would be a great loss, I would feel the loss of those. See that tree out there? It's an Australian red cedar, thirty years old, because I planted it. It's going to be hard to leave that.*

Jenny Lewis: *[For me it's] just the totality of it and the beauty and the pleasantness of the whole valley. You can go down there and you can't hear a thing.*[35]

Before the house was completed, in 1966 Ivan Lewis received notice of the planned county road. 'Advice has been received from the Department of Main Roads that the Expressway is at this stage a tentative line on a map. The proposed route had not been pegged as yet.'[36] A Department of Main Roads map showed a dotted line winding ambiguously down Devlins Creek. An urgent meeting of local residents was held at the wooden footbridge. Sir Garfield Barwick, still the owner of a house on the estate, addressed the meeting. A petition was sent to premier Lewis, but still no one could discover exactly where the road was to go. For several years nothing more was heard from the traffic planners. From the later 1960s it seemed that the plan had been forgotten.

Orchard Road lies between Mahers Road and Lynbrae Avenue. To this street Julia Garnett and her husband Tony moved in 1966.

We were the first house in the area. Across the road was all bush ... We really felt like pioneers then, because the roads were just dirt tracks. I remember my father said to me, 'Why do you want to go and live out there in the bush?' It did seem like out in the bush. We used to come and visit the block before we built here, and when we were the first house here we were surrounded by bush. A couple of times we got bogged where the creek flowed across the road. And Mahers Road, the trees grew up the middle of it, so you weaved in and out of the trees by following the tyre tracks. Then [the houses] were built one by one.

We've had a wonderful little community here because we all had kids at the same time virtually. They had three next door and they had three [on the other side] and we had two at that stage, and the kids formed a little gang and they lived in everybody's houses, so it was open house to all the kids, and we had a swimming pool, they all learnt to swim in the pool. Next door had a trampoline and they all learnt to do back-flips on the trampoline, and next door that side had a swing set and a barbecue, so we regularly had a neighbourhood barbecue in there, swimming in here. It was lovely. It was a great place to bring up kids. With the bush there, we'd often follow the creek down, and the kids would play in the creek. It was just beautiful. When we first came, the people next door said there was a platypus in the creek. But the water board put a sewer line right along the creek and it became polluted, I guess that was one of the sad things about civilisation.[37]

Despite occasional worrying rumours, the county road proposal remained dormant until 1988. Then abruptly the Roads and Traffic Authority announced its intention to build not the minor county road but a concrete freeway:

> Planned population expansion in north-west Sydney will add to congestion on the east–west road corridor through Epping. A freeway is proposed to improve travel on the corridor and reduce the adverse impact of through traffic on local streets. A solution which provides a balanced road network must also consider social, environmental and economic objectives.[38]

The Castlereagh Freeway, from Pennant Hills Road to the Lane Cove River, known as the F2, would be a four-lane dual carriageway with emergency shoulders and an exit ramp near Mahers Road. What would it mean to the residents? When would construction start? All that was clear was that the proposal involved a sweeping carriageway constructed over and through the northern slope of Devlins Creek. The Pennant Hills Golf Club would lose 20 m of its southern boundary. Esme Blackmore would be eligible to sell her house to the Authority because the expressway would pass outside her front door. Julia Garnett would be some 20 m from it, and Ivan and Jenny Lewis about 100 m. The prime casualty would be the bushland; the northern slope of Devlins Creek was the best preserved and least affected by exotic and noxious weeds. A bus hired by the Authority took the Lewises and other residents for a drive up a completed freeway to demonstrate how quiet and well landscaped road construction could be.

Planners of major freeways are required under New South Wales law to make an environmental impact assessment of any project. It must consider the description and justification of the project, relate it to other road systems, assess the alternatives, the loss of urban bushland and open space, the effects of noise, air pollution and lighting on existing residential amenities, the likely impact on existing transport, and safeguards concerning drainage and the ecosystem.[39] There is almost nothing in the Act requiring an environmental impact statement to consider the attachments which individuals may feel to their own particular sites. Bushland may be protected by law for its own good or because it fulfils a public purpose, but not because individuals feel attached to it. Affection, memory and experience have no permitted place in the assessment. The Act directs assessors only to consider the 'need for any resumption and/or demolition of buildings'. No more. Most submissions to environmental impact assessors therefore stress the generalised, representative nature of bush or amenity, not its meaning to the individuals who will see their neighbourhood destroyed.

The Roads and Traffic Authority carried out its own environmental impact statement on the proposed route of the F2. It found that the freeway was 'clearly justified in terms of the economic, transportation, social and planning objectives'. Although there would be adverse environmental effects, it proposed 'suitable safeguard measures'. 'It is considered that construction of the freeway is justified overall.'[40]

Intense protest at the predictable in-house findings centred on the need to retain the bushland. The Friends of Beecroft Reserve organised T-shirts ('Treeways not Freeways') and rallies. Seven hundred people gathered on 9 July 1989 to hear politicians, an economist, a traffic authority and a student speak against the proposal.[41] So trenchant was local and press opposition that the Roads and Traffic Authority appointed a commission, headed by the chair of the Commissioners of Inquiry for Environment and Planning, to reconsider it. The Woodward Commission began hearings in October 1989, six months after the first (now discredited) environmental impact statement had been released.

The Commissioner conducted eight months of hearings involving 829 submissions from 301 parties. The proposed access road to the freeway would affect Mahers Road, and the residents, including Esme Blackmore, formed their own lobby group to oppose it. She wrote a submission citing disturbance, the destruction of bushland and the state government's broken promises. She took Commissioner Woodward down the street and into the bush to demonstrate how the neighbourhood would be affected. *I told him how I get to a certain tree line returning from Sydney, and how I get this greatest feeling of satisfaction and peace looking out onto this little valley.* Ivan and Jenny Lewis pointed out that there were many more species of native animals and plants in Devlins Creek than had been counted in the Authority's rather casual assessment. The community calculated, however, that the better strategy lay in stressing environmental problems rather than the destruction of the neighbourhood. Woodward was impressed. His report noted great concern at the likely traffic noise levels, particularly at night, the significant visual impact of viaducts, embankments and cuttings, the degradation of the atmosphere, the substantial loss of residential amenity along the route, and the destruction of a large area of urban bushland. He found that the environmental impact on several hundred residential properties would be far worse than those envisaged by the Authority's environmental impact statement. The 'current leafy outlook' from residences to the golf course would be substantially changed. Existing views would be replaced by the vertical walls of inadequate sound baffles. In sum, the Commissioner found that 'Social impacts on residents are likely to be substantial. The social impact analysis undertaken by the RTA fails to establish that social benefits outweigh adverse social impacts.' Woodward's recommendation was that the freeway not be constructed.[42]

The Commissioner's terms of reference were similar to those of a standard environmental impact statement. He was not instructed to consider the attachment of individuals to particular homes or suburbs and, except when specifically informed by verbal submission, did not. Nor did the residents offer much of their private feeling to public analysis, partly through the public nature of the hearings, and partly through the tactic of stressing potential noise and pollution. Some 100 buildings were inspected by the Commissioner—for their heritage or architectural value alone. No resident was consulted about their *feelings* about the loss of a building and, as it happened, no building on the proposed freeway route was on the register of the National Estate. No house in Mahers Road was 'considered to have any special architectural significance'.[43]

In 1994 the Devlins Creek bushland was extensive, serene, mysterious and, for Sydney, remarkably unaltered.[44] Half a kilometre wide and 1.5 km long, a visitor heard nothing of Sydney traffic; it was easy to sense what it was that western Beecroft residents had valued so highly. The official assessors walked the area and considered its value in language of singular contrast to that of the Lake Pedder conservationists. While the Pedder protesters savoured a 'healing place', a 'sanctuary of profound beauty' and a 'temple', the language of the Authority's environmental impact statement was almost entirely negative. Conceding some potentially harmful effects on the environment, it referred generally to the lack of introduced vegetation of any cultural significance along the route. It identified only remnant pockets of native bushland, for exotic trees like willows had so infiltrated the bushland so that only three areas remained in good condition. The discharge of urban storm water added to the burden of weed invasion. It granted that substantial areas of Devlins Creek remained in a viable condition, but the invasion of exotic plants had followed each disturbance, so that native species had been affected; the tall canopy of eucalypts would be gradually replaced.[45] Woodward's findings were much more positive, but he too seemed to have been locked in a linguistic prison from which poetic analogies were debarred. His appreciation was limited to 'leafy', 'enjoy', 'open backyards', 'significant', 'good quality'. Devlins Creek was 'a significant remnant', the bushland was a 'significant visual and natural element', 'visually dramatic', 'visually attractive', an 'attractive contrast', a 'natural contrast', a 'relaxing containment', 'quiet and pleasant', 'quiet and stable' whose destruction would be 'visually unacceptable'.[46]

Woodward's discussion of community feelings about the project, which reflected the terms of the submissions put to him by the residents, was limited to the terms of settlement, eligibility for purchase, method of valuation, community awareness and the effects of continued uncertainty.[47] He noted that the residents whose houses were ineligible for purchase had objected to noise, unsightliness, the necessity for avoiding verandahs and closing certain windows permanently. Devlins Creek, the significant remnant, was no healing place, temple or sanctuary; those terms were confined to wilderness arbitrarily defined in contemporary culture as remote, undisturbed and 'a legacy of the beginning of life on Earth'.[48] The Commissioner judged the bushland to be not wilderness but an attractive contrast, corrupted, invaded and proximate. He wrote not a word on the particularity of loved home, garden, street, suburb or bushland.

Residents' celebrations that the F2 should not proceed were tempered with anxiety, for Commissioner Woodward had recommended that the project be reconsidered in a decade. Would the Authority follow his advice? The worst fears soon were realised. Only a few months after Woodward had rejected the freeway, the Authority reinstated it, this time as a tollway following a slightly different route, and with an additional 12 km pushing towards Blacktown from the other side of Pennant Hills Road. The altered track took the freeway some 40 m further south. The golf course would now be unaffected, but the Lewises would have the freeway 20 m nearer their house. All the dwellings in Mahers Road, including Esme Blackmore's, were in the path of the revised freeway. So were the top two houses of Orchard Road, including Julia Garnett's.

In her submission to the Woodward Commission entitled 'Bushland at Work', Julia Garnett described Devlins Creek as a 'green way', part of the network of state-wide green corridors to help the survival of birds and other wildlife. The Authority had distorted its environmental impact statement, she argued, to present Devlins Creek as a weed-infested deteriorating remnant. Not so: the Friends of Beecroft Reserve were gradually clearing exotic species from the bushland. She quoted a World Bank economist on the folly of regarding the environment as a limitless resource and dumping-ground. She cited the ability of trees to raise water to the surface, control erosion, cool and clean the atmosphere, and affect the climate: 'Scientists tell us we have "probably ten years" in which to turn around our presently destructive lifestyle to preserve a future for our children and our grandchildren. We need more trees. Fast.'[49] In this impassioned and mostly generalised submission, Julia Garnett did not touch upon what was of equal importance to her—the personal and intimate specialness of the particular bushland which surrounded her.

Julia Garnett's Beecroft was a suburb-within-a-suburb, ten houses down Orchard Road, ten houses up Mahers Road, ten more in other streets, the family home itself, the familiarity and attachment to the verandah and the lounge where thousands of cups of tea had been drunk. Yet most of Julia Garnett's special spots in her home and suburb were special trees or plants. She and her husband Tony planted a bluegum when one child was born, a lemon-scented gum for the second and a magnolia for the third: *That's a bit traumatic when you think they're going to be just bulldozed.* Except for a turpentine and a single bluegum left undisturbed in the old orchard, they had planted the entire garden themselves, landscaped the beds and the lawn and laid the bricks. *It's a bit devastating to think they're going to be all under a freeway.*

Just behind the Garnetts' swimming pool is an old turpentine tree:

I'm more upset about that tree, I guess, than the house itself. Because I've lived next to that tree for thirty years, and it's just been a wonderful experience. If ever a tree had a spirit or an aura about it, that tree does. It's full of birds, I think it must be at least a couple of hundred years old because the the other turpentines which are quite small, barely changed at all. And this big tree has remained the same in the whole thirty years we've been here ... Knowing just how slow-growing it is, it's just ancient. It probably saw Aborigines here. I just have a feeling that it's such a presence, it's seen so much and endured so much, and to think that it'll quite likely be just chopped down is, I don't know, sacrilege. I'm an artist, and I've drawn it and painted it so many times. You really become familiar with something when you draw it. I guess it means quite a lot mean to me, that tree ... I think about it quite a lot.

The tree always seemed to her a friendly presence. Whenever she swam in the pool she looked up through it. *It's just so beautiful.* To the question, 'Does the tree know when you're there?' Julia Garnett replied, *Well I wouldn't say that it doesn't.*

Esme Blackmore's neighbourhood included the Pennant Hills high school where she taught, the bush where all the families had picnics, the Scout and Guide halls and the West Epping shopping centre. She too had spoken at the angry meetings of angry residents and, like Julia Garnett, found that what was most important to her was too intimate to be revealed in a public report.

While Esme Blackmore spoke of moving out of her home in Mahers Road she also spoke of when she had moved in. She and her husband had watched it being built, overseen the builders, chosen the tiles. Their architect friend had taken so much care over the design and construction, and had even directed the workmen to rebuild the fireplace because it did not match his exacting standards. *I can still see him doing that now.* Esme Blackmore had been ill with appendicitis just before they had moved from Hunters Hill. Friends helped carry things in on the day. It seemed like life had begun again.

For this was the only house that she and her husband had owned together. John Blackmore read dozens of books on garden and pool design; in those *maddening and joyous* years, dozens of rocks large and small were brought in and arranged. Certain plants in the garden had been given by clients, a camellia by a relative to mark a special occasion, a peony by a friend going into a nursing home, house-warming plants, special anniversary plants. *I can't replace that garden.* To and from the house flowed the activities of the family: rushing to Guides in one direction and to Scouts in another, out somewhere every second night. *I was rushing back and forth, you complain at the time, but it's all part of the excitement.*

The most special places were inside the house, in the carefully designed family room where the children, *unfettered and uncriticised,* played, read and listened. They pulled their toys from a wicker box kept in the corner, learnt to crawl and took their first steps. *It will be sad when it is pulled down and reduced to a pile of rubble.* Here the family took their meals; everyone talked; the children learnt *to get a point of view across to their father, they had to say it more loudly, so there were great shouting matches across the table.* Had the walls absorbed some of the atmosphere of those joyous years? *Well I think they have.* Did the house know when Esme Blackmore was in it? *Yes it talks to me, but it doesn't like to be neglected, this house. It says 'Get the vacuum cleaner and get the dust off me.'* Her home made Esme Blackmore feel *safe and comfortable*; it was *filled with familiar things;* it made her *quite warm.* She hoped that another family would occupy it when the time came for her to leave. *I'm not going to see this any more.*

Memories were mixed. John Blackmore died after only ten years' residence. Another phase of life began when the children married. Ten relatives-in-law arrived to stay for the Royal Easter Show. *This happy home.* Esme Blackmore's son and his wife lived in the downstairs flat; the three got on well together, shared the housework and shopping; it was a *most joyful time. I think of them in the flat below and I'm sad to think of it as empty and won't be used for anything.* Like Margaret Johnson, Esme Blackmore delighted in her grandchildren reading the same books as her children. They played with the same games from the same wicker box in the same corner of the same family room. Esme Blackmore watched one of them, *my poor little two-year-old grandson* die slowly before her eyes of adrenal gland cancer as she nursed him in the kitchen. One day she had gently led him away from the power point that his fingers were straying near. *I can see this lovely little golden-haired boy, a vision I have, a lovely memory.* There, in the family room, by that power point, in that corner. That place. When she felt depressed away from her home, she thought of its particular rooms; when at home she sat in a particular chair in the lounge, her husband came to her. There, in the lounge, in that chair.

The Authority did not proceed on its revised and extended construction without a further storm of environmental and economic criticism. In 1993 its opponents reasoned, as had Commissioner Woodward, that a tollway could not be cost-effective. The sustained campaign which caused the earlier route to be changed was redirected at Authority calculations that the F2 would reduce traffic on roads beside it or at either end. 'The F2 will not bring Sydney nearer to some imagined ideal of uncongested roads ... The F2 will not take travellers where they need or want to go.' In addition to 'devastating the built and natural environment, the tollway would suck an estimated $600 million from public and private funding'.[50] John Goldberg of the Department of Architectural and Design Science at the University of Sydney surmised that the real cost would be closer to $650 million than the official estimate of $400 million. He claimed that the Roads and Traffic Authority had withheld essential information from the Woodward Commission and that the traffic volume projections were deliberately inflated. Despite the provision of two bus lanes, there would be no opportunity for residents along the 11 km of freeway to catch one. He calculated that a much larger number of residents than estimated would have their sleep affected by traffic noise, and concluded that the Authority was guilty of bureaucratic misfeasance (the wrongful exercise of lawful authority).[51]

The Authority proceeded as if public funding and government support were guaranteed. Julia Garnett and Esme Blackmore were visited by valuers, who were to add an additional $20 000 to the Authority's initial valuation of residences which had been occupied by the same people since 1965, the year of the county road plan. The public were invited to liaise with the Authority's two Community and Environment Officers: 'Feel free to contact Kerry and Bill during office hours if you need some information regarding the impact of the project or the community impact of its design.'[52]

Still nobody was coming to grips with the grief and loss soon to be sustained by the evicted residents and those who loved the bush. Among the millions of words which had been expended for and against the project, only two sentences in a special supplement to the journal *Urban Action* came close to appreciating the grief some residents were feeling for their dying homes: 'A fair price for brick and mortar is better than an unfair price, but money can't buy the love, energy and the time which families invest in homes and communities. It can't buy the confidence and strength to start again.'[53]

Start again, though, they must. Ivan and Jenny Lewis cannot see themselves remaining. Though friendly with their neighbours, their suburb changed after the children went to school, from the streets and shops, parks and swings of a young family's mental map, to the bushland into which their backyard merges. Their social life has for six years revolved about the Coalition of Transport Action Groups: *They are my life.* The Lewises have fallen out with some members of the Beecroft Cheltenham Civic Trust who, living away from the planned route, see traffic being drawn away from their own roads. *They consider us 'nouveau Beecroft'—and are expendable.* Though not directly in the path of the expressway, the Lewises' continuing residence will be intolerable: Goldberg demonstrated that houses with open windows facing an expressway fitted with roadside noise barriers can be located no closer than 200–300 m from it;[54] the Lewises' house will be situated some 60 m from the F2 East. Ivan Lewis contemplates his years of work in the garden. *I regret having done it at all now.* Jenny Lewis already found her Beecroft a car-orientated dormitory suburb, but she loves the bush and headed the local regeneration group. She was obliged to set it aside as their home became the headquarters of the Save Devlins Creek campaign. They reflect upon the brave promises of the 1993 Sydney bushfire victims to return and rebuild. *They can't return here, it's under concrete.* They see no psychic dimension to their occupancy: *Not really, but the birds know I'm back. I'm not spiritual, but the bush has improved.* What, in the end, were they fighting for? Jenny Lewis answered *For the bush—it has a right to exist too.* Ivan Lewis answered *My footprints are here.* The hoop pine at the corner of Ivan and Jenny Lewis' garden stood 35 m from the route of the freeway, but the Authority did not grant them an opportunity to sell.

Throughout 1994 Julia Garnett's suburb was vanishing piece by piece. One hundred and forty-five home-owners along the route asked the Roads and Traffic Authority to begin negotiations to purchase their homes. They seemed to include all her neighbours. *For a while the moving van seemed to be coming in every week.* The houses of her old friends lay derelict for a few weeks or months, then were filled with young families on short-term rentals from the Authority. *It's nice to hear the kids again but you can see the houses start to deteriorate. What used to be beautiful lawns are all dry and weedy. Pride in the gardens is the first thing that goes.*

What would the Garnetts take with them and where would they go? Julia Garnett planned to take all the bulbs, *tons of things I've left in pots, if we can manage the physical work in all the moving trauma.* She preferred a house, not a unit, at Longueville, while her husband hoped for one near Darling Harbour. The Garnetts would have to find a place suitable for their son to continue university studies; Julia's art teaching would have to end if the family moved far away from her school. Her daughter felt a melancholy satisfaction that their family alone lived in their home from its birth to its death. *Like a funeral pyre.* With so many friends and supporters already gone, Julia Garnett felt *nervous about publicly fighting when the RTA is buying your house.* Did the valuers underestimate home-owners' attachment to their houses? *Oh no, it gets no value.* Her family probably would be the last to move from Orchard Road. *Everyone is now looking for new suburbs.* The turpentine tree behind the pool symbolised to her all that had been most precious to her and to all the previous generations who had shared the site:

The Turpentine Tree

Friend of birds
Laden with blossom
Knew the Aborigines
And the Settlers
The Soldiers and the Farmers
The young couple
Who built a house
and raised a family
Knew them all.

Sacred Tree, Tree of Mystery and Life
Dance by moonlight
Stars in your branches
Dance on my friend
Dance on.[55]

Esme Blackmore's first notice of compulsory purchase arrived in June 1993. The white company cars of the environmental impact assessors, the Roads and Traffic Authority and sundry investigators seemed thicker on the roads week by week. A real estate agent arrived to offer her a red-textured house in the next street. *How insensitive. Where I could see and hear the bulldozers destroying this house. I can't believe it!* Three other agents arrived in the same week. A state government valuer came unannounced and left his card. *Very unprofessional. I wouldn't have let him in.* Her next-door neighbours, both of whom had arrived in Mahers Road before the Blackmores, sold and left. One neighbouring couple were so attached to their property that when the wife died, her husband scattered her ashes on the front garden. His moving van arrived. *Terrible.* The house lay empty for several weeks. *Unnerving. Unnerving.* A number of young people took out a short-term rental from the Roads and Traffic Authority. The first Esme Blackmore knew of them was the mysterious arrival of a large pink unregistered bus on the front lawn. A few weeks later mystery turned to bizarre comedy: a noisy party was still proceeding at 6 a.m. when Esme Blackmore finally remonstrated with what appeared to be groups of Valkyries sitting around little fires on the back lawn. A warrior courteously removed his helmet and agreed that their neighbour deserved a fair go. On occasions nine cars were parked on the front lawn, once so meticulously ordered and where the ashes had been scattered of a woman who loved it. *It's a wonder she doesn't get up and do something to their differentials.* Some of Esme Blackmore's neighbours have gone to the north shore, the Blue Mountains, or to the new and bare suburb of Cherrybrook. *I hate Cherrybrook.* The neighbours down the hill rented out their home and went to live at Lismore. The old community was disintegrating. Esme Blackmore lunched with some of her former neighbours now scattered around northern Sydney in flats and units. *They're a quarter of the size of this.* She planned to move when the last of her friends, the Garnetts, went, probably by the end of the year. *Remaining here is sweet and sour.*

Esme Blackmore made an offer on a house in the north shore suburb of Turramurra. The spiral stairs, the tallow floors would have to remain in Beecroft. *Just to scrunch them up is appalling.* She would have liked to take the whole garden, rocks and all, especially the boulder by the front gate, bought thirty years ago, rolled in by a front-end loader with great ceremony and now covered in lichen. And all the plants. *Where am I going to get trees and shrubs like this?* She wanted to take the floor-to-ceiling cupboards she designed long ago, the extra-large laundry. *I can't replace them.* She would keep the pictures and the piano even though nobody played it much now, and the hundreds of books. *I have to have the books there, they're part of my life.* Would she return? For bush regeneration, yes, on the other side of the creek, but never to live in this suburb, never to return to this place.

I just feel angry that governments and politicians have let us all down and destroyed our life. And the people that seem to have power through companies they work for seem to be able to push ordinary mortals around. I feel powerless against all these people whose company cars seem to give them the power to say 'We have to have a road because we want to drive our company car hither and yon and we don't care what happens to your house.' Even our newly elected member of parliament keeps putting little dodgers in my letter box saying 'The people of Baulkham Hills are very pleased that the M2 is about to be built.' And that goes into my letterbox here!! He doesn't say, 'And I'm very sorry that your house has to be sacrificed to do it.' I'm really angry that in a mysterious redistribution of seats, this little area following the route of the Freeway was hived off the previous electorate that it was in ... I just feel that we've been pushed around for the sake of the motor car, and for the sake of people who don't even have to buy their cars or pay for their cars or put petrol in them.[56]

October 1994 was an agonising month. All the remaining residents of Mahers Road had been asked by the Roads and Traffic Authority to sign the deed of sale of their homes by the end of the first week of November 1994. Already many neighbours had gone and their homes were vanishing. Twenty-five years earlier Julia Garnett had watched the sites of their neighbours' houses being cleared, then the homes erected; now one by one she watched them smashed. One morning she left for work as the demolishers began work on the house next door, carelessly dropping and breaking the tiles from the roof frame. *The waste of it all.* By the end of the first day the roof was gone. By the end of the second the interior was smashed by a machine which looked like a small Yallourn coal-dredger. *A big cruncher just biting away at the walls.* The curtains and blinds, which the Authority had ordered should remain in the house, were pushed around in the dirt by a bulldozer along with everything else. By the end of the third day only a pile of rubble remained, and by the end of the fourth that too was gone. In the far corner of the block a fragment of lawn, half a metre of stone border and some geraniums remained untouched. The tenants quit the house where Julia Garnett's friends had once held barbecues around the swing set. She invited them to collect any last plants or souvenirs; two weeks later the demolishers had not yet arrived. At night, only silence and darkness.

Empty houses opposite the golf course. Something odd settled about homes within a week of the removal van's departure: the drawn curtain, the missing shrub, the wilting plant, the fallen leaf, the lengthening grass, the detached flyscreen. Half-demolished houses at the end of Mahers Road: muddy carpet, cracked tile, naked wall, holed window, gaping fibro, fallen batts, copper stovepipe on the lawn, a kitchen stove amid the wreckage of the kitchen. Uncleared sites in

Ferndale Road: half-a-dozen driveways lead to a demolisher's cyclone-wire fence, raw white sandstone sharp on weathered grey; among piles of unrecognisable rubbish lie a white letterbox for numbers 14 and 14a, a triangular sign identifying number 12, a raffia mat, fragments of retaining walls, half a staircase, a garden light illuminating no path, a window frame, air conditioning ducts, bits of trellis; down by the creek lie three smashed and empty swimming pools. Cleared sites at the top of Orchard Road: tiny fragments of attachment—a peg, a washing-up brush, corners of garden beds, the end of a hose, a tennis ball, half a pool sweeper, a garden rake leaning on what was two weeks ago a back fence. Stranded objects. The destruction of the neighbourhood seemed an analogy of the fragility of human life itself: what seemed so strong, so permanent was snuffed out in an instant.

When I saw Esme Blackmore for the last visit to her home in Mahers Road she pulled back the curtains. *I'll take a last look at the golf course.* A nearby resident holding out against the Authority's valuation dropped in. *When are you going to have your garage sale?* The two neighbours surmised that the community, now scattered over a third of Sydney, would fall apart. *We're the only ones left. We've gotta stick together.* There seemed little hope of remaining in their homes after Christmas. *A road that's going nowhere. It'll look like Hiroshima when they've finished.* No professional had visited Mahers Road, Orchard Road, Lynbrae Avenue or Ferndale Road to counsel the exiles in their distress, no local or state political representative, no official save a valuer. *They treat us like lepers.*

In December 1994 the removalists came for the Blackmores. Esme's son David photographed the house and garden. She and her friends dug up the Japanese maple, the peony, the macrazamia, the azaleas and a few of the sandstone rocks. The next day the larger trees remained half dug out; shrubs had been trampled by the removalists and the electricity meter had been taken away. A few return visits were necessary for the remaining plants but nobody much wanted to go. *It's dirty and empty. Desolate.* Every artefact remaining in the garden was fast becoming, as in Macedon, a strange and stranded object. The leaves settled and the grass grew longer.

CHAPTER 9 | THAT PLACE

Are local places important? How can assessors or friends gauge the significance of a place to people who may scarcely realise its value to them until that place is threatened? How, when a loved place is destroyed, should dispossessed people receive emotional support?

The concrete road which destroyed west Beecroft is a phenomenon of creeping international sameness, the victory of 'universal civilisation, symbolised by the serpentine freeway and the free-standing high-rise office tower, over locally inflected culture'. The bulldozer has been described as the 'agent of *tabula rasa* modernization, the technocratic gesture aspiring to a condition of absolute placelessness'.[1] We have watched this tendency of Euro-American culture towards delocalising and devaluing the built landscape in many places: in eighteenth-century Britain, at Adaminaby, Jindabyne, Cribb Island, Yallourn, Weetangerra, Goulburn, Darwin and now Beecroft. For several decades scholars have pondered the implications of the continued draining of particularity from regions and cultures and even from nations.[2] Are we doomed to a uniform 'megalopolis', as one critic put it, proliferating until internationally inspired planners reduce cities to 'little more than the allocation of land use and logistics of distribution?'[3]

Not only cities suffer from creeping sameness, so do towns and suburbs. Cities like Darwin and Sydney can dominate as well as be dominated. We have seen bodies like the Sydney and Melbourne City Councils destroying Surry Hills and Carlton. Sydney, Melbourne and Darwin are individual urban cultures, but so are the individual suburban cultures which together form those cities. The destruction of old

Surry Hills or west Beecroft may be as much a loss to Sydney as the loss of locally inflected Darwin would have been to Australia. Why preserve Darwin, Beecroft, Cribb Island or Adaminaby? Because they may be the creative nucleii of future cultures. Because localities should not signify the boring or the provincial but 'subjectivities with their own integrity'.[4] Because an ordinary home or suburb, once loved, transcends particularity for as long as the person who formed the attachment values that bond. Mystery and pathos, as Douglas Stewart put it, seep from earth and bubble out from water in any place where people have loved and bred and feuded with each other. Each construction of event, experience, memory and place is unrepeatable.

Lastly, loved sites are worth preserving because of the intense pain which their destruction may cause to the inhabitants of those places. Post-cyclone Darwin, Mahers Road in Beecroft and Macedon town were each compared by uprooted residents to Hiroshima. Yet not a single person in any of the accounts we have followed received any kind of counselling for the grief and trauma associated with their lost place. Many received, instead, the unsolicited advice to get on with their lives. Adaminaby residents were encouraged not to mourn their drowned gardens in the old town but to compete for the 'best garden' prize in the new. Polish Holocaust survivors did not always wish to speak about their experiences because of the unspoken question: what must you have done to stay alive while millions died?[5] The Australian National Parks and Wildlife Service predicted that 'it should be possible to arrange alternative accommodation for those residents [like Granville Crawford] without undue hardship or great expense'.[6] Three months after Cyclone Tracy, senior health counsellors ordered that post-trauma counselling to victims cease, because attention would 'only emphasize the problems and initiate or perpetuate psychological distress'.[7] The Roads and Traffic Authority felt obliged to do no more than invite residents of Mahers Road Beecroft to liaise with the Authority's two Community and Environment Officers: 'Feel free to contact Kerry and Bill during office hours if you need some information regarding the impact of the project or the community impact of its design.'[8]

Australia is not alone in ignoring the psychological effects of place deprivation. Environmental psychologist Gerda Speller found, in her studies which involved destroyed or resumed English villages, that not a single affected person had been offered counselling for the grief they felt through loss of place, nor were their psychological needs ever considered to be an issue by the authorities. As in Australia, environmental impact assessments had considered the usual criteria of dust, noise, vibration and environmental damage. They had not, however, assessed the impact of loss of home, community and countryside, even though Article 3 of the EEC Council Directive governing assessment procedures directed attention to the direct and indirect effects of a project on human beings, water and flora. A correspondent told her, 'my brother comes to see me. Well, he doesn't really come to see me, he comes because he likes to come home. He won't be able to do that once this house is demolished, there will be nothing for him to return to.'[9] Speller concluded 'Rather than deny that a problem exists, we need to articulate it and find ways of building a supportive structure.'[10]

Perhaps everyone who has made the journey to nothing knows that longing which is both collective and lonely. 'To watch another suffer', as Anne Lindbergh knew well, 'is to know the barrier that shuts each of us away by himself. Only individuals can suffer.'[11] Grief for dead places seems much more analogous to grief for dead people than professional carers have allowed. Cultural studies, the dominant mode of analysis of contemporary society for the past two decades, has shown little interest in the personal and individual. We need a second Elizabeth Kubler-Ross to advance place-bereavement as a continuing theme of contemporary distress.

It is now time for environmental and heritage assessments to encompass these profound emotions. The 1974 Inquiry into the National Estate, while recommending that residents should be consulted before redevelopment began, put its faith in conserving the best features of the built environment.[12] Since then heritage assessment criteria commendably have sought greater input by the community to determine which of our collective memories and physical valuations should be preserved. Through such instruments as the Burra Charter, the Australian Heritage Commission and other bodies have rescued hundreds of significant buildings, sites and complexes from destruction or decay.

The guidelines by which an assessment of preservability is reached are complex. The current criteria for the registration of a place in the National Estate look at the totality of aesthetic, religious, spiritual, symbolic and educational values which are placed upon a site by experts together with 'a' (not 'the') community. The specialness of a place to an individual is not normally considered except where that individual is 'of importance in Australia's natural or cultural history'.[13]

The saved places are undoubtedly significant, but are they held in affection? Any emotions felt by individuals for threatened sites are currently held to be tangential to the more quantifiable yardsticks of national/regional importance or community valuation. Yallourn excepted, probably none of the built environments visited in this book would have qualified even for listing in the National Estate because —to the outsider—the same kind of town or suburb or house could have been found anywhere. The residents of Yallourn undoubtedly wanted their town saved, but a majority of the community of the Latrobe Valley might have have felt that Yallourn could be destroyed, especially if its prized features were to be redistributed among neighbouring towns.

Community valuation, then, is a slippery term. If Lake Pedder was nearly saved because conservationists were able to universalise the values of wilderness, twenty years later it is time to seek out and listen to the voice of individuals within whatever one takes to be the collective voice of a community. Clearly no community is a single entity. In the 1980s 100 people assisted the bush regeneration of Devlins Creek. In 1995 100 more people opposed the bulldozers as they clear-felled the route of the M2 through the same area. The west Beecroft community facing extinction was all the people of Mahers Road *and* those living in Lynbrae Avenue and Orchard Road, *and* other residents of threatened streets along Devlins Creek who became known to each other as the campaign proceeded, *and* the smaller clusters of neighbouring people in Mahers Road who knew each other well before the threat, *and* the residents of houses and gardens of the individual families whose homes had lived and died within a single generation. The threatened community as a whole did not agree on whether saving the houses or the bush was the first priority. Outside the threatened community, other citizens of Beecroft were not necessarily opposed to the M2 at all. The residents—those insiders who made up the 'real' community—identified and clustered in a number of ways, while outsiders, including the environmental impact assessors, took them as a single entity. If cities can both be threatened and oppress their own constituents, so can suburbs, neighbourhoods and communities.

A community, then, is multi-factored and multi-purposed, and understanding its components depends partly on whether one is looking out or looking in, that is, whether one is or is not a member of the threatened group. Heritage and environmental impact assessors have not yet been able to appreciate this multitude of valuations among insiders who look out towards the threat. This should not be surprising, not because such assessments are so difficult, but because assessors have not yet seen the need to appreciate the valuation of the individual, the family, the neighbourhood, the suburb and the town which coexist within the 'community'.

Not all neighbourhoods or houses can or should be saved, or the living generation will be crushed by an intolerable burden of history. Driving from Canberra to Beecroft to continue work among the threatened community, I travelled on another new concrete freeway near Liverpool which cut twenty minutes from the previous journey. No doubt for this construction also, dozens of people had been evicted from their homes which had then been demolished and forgotten by outsiders. Of such dispossessed people I knew nothing, as others knew nothing of the private griefs of the residents of Mahers Road. Such feelings of residents for their threatened loved houses or towns can never, perhaps, be more than a single factor in any environmental impact or heritage statement, *but at present they are not represented at all.*

Memories are ghosts that won't lie down. Susan Karas was reluctant to return to the family cemetery at Zilina because evil had contaminated the whole of her Czechoslovakian birthland; Arnold Zable could not understand these same events until he located and stood before the sites of wickedness. Alan Lucas and Eleanor Gray hold keepsakes close to remind them of Yallourn and Macedon, while Granville Crawford keeps only his memories of the high country. Alison Rawe neither wants nor needs to see her family's home in Shadforth Street, Mosman. In 1995 she and her husband sold their Cremorne home, with which they felt only fragile association and moved to Mittagong where John Rawe had spent his childhood.[14] Eleanor Gray and thousands of Darwin residents felt the need to be close to their lost sites during rebuilding. Margaret Johnson returns to the land, but not always to the homestead, of Windermere Station.

If memories are unlaid ghosts, the physical past returns to haunt those who thought they had completed the journey to nothing. Twenty years after its inundation an underwater camera captured the tyre-marks of light aircraft on the floor of Lake Pedder.[15] In 1995 the worst drought for fifty years dropped the level of Eucumbene Dam 25 m below highwater mark. Pieces of stove, sulky, bed and motor car emerged among the foundations of houses so scattered and smashed in the site of the old town that they seemed to have been razed by an invading army. Nita Stewart stared at the naked forms of the dead pines revealed above her grandmother's home and marvelled how deep the water normally towered above the sites of her lost childhood. In Darwin Suzanne Parry was astonished to find that every wet season for six or seven years after the cyclone, a new crop of 1c and 2c coins would appear in the same spot on the lawn of her new home.[16] Anne Boyd, in her rebuilt home in Ferntree Gully, was content: *as summer storm clouds blow in from the west, I seem to feel ever more passionately about the new house.*[17] In April 1995 twenty people, some perched at the tops of trees in Devlins Creek, were arrested for obstructing progress on the M2.[18] Esme Blackmore's house had vanished. Julia Garnett's house, its windows guarded by steel grilles, had become a construction site office. She was assured that the turpentine tree would be saved, but she remains doubtful.[19]

Prue McGoldrick, holidaying on the Isle of Skye, smelt the scent of peat fires reminiscent of the brown coal of Yallourn and thought of the words of Proust:

> When from the long distant past nothing subsists,
> after the people are dead,
> after the things are broken and scattered,
> still, alone, more fragile, but with more vitality,
> more unsubstantial, more persistent, more faithful,
> the smell and taste of things remain poised for a long time,
> like souls, ready to remind us,
> waiting and hoping for their moment,
> amid the ruins of all the rest,
> and beat unfalteringly
> in the almost impalpable drop of their essence,
> the vast structure of recollection.[20]

NOTES

PREFACE

1 Philip Hodgins, 'On Going Back for a Look', in *Down the Lake with Half a Chook*, ABC Publications, Sydney, 1988.
2 See below, ch. 7.
3 Dorothy Hewett, 'In Summer', in *A Tremendous World in her Head*, Dangaroo Press, Sydney, 1989, pp. 94–5.
4 Laura Neal (ed.), *It Doesn't Snow Like It Used To*, NSW Govt Printer, 1988.
5 M. Mackay, 'Adaminaby The Old Town', in ibid., p. 17.
6 Douglas Stewart, 'Farewell to Jindabyne', in *Collected Poems* 1936–1967, Angus & Robertson, Sydney, 1967, p. 28.
7 Peter Read, 'Remembering Dead Places', *The Public Historian*, forthcoming; 'My Footprints Are Here', *Oral History Association of Australia Journal*, 17 (1995); 'Lost Places and the Language of Destruction', *Australian Folklore*, 10 (1995), pp. 160–7.

1 LOSING WINDERMERE STATION

1 The narrative of events at Windermere Station is drawn from interviews with Margaret Johnson and Jim Johnson.
2 Gaston Bachelard, *The Poetics of Space*, tr. Maria Jolas, Beacon Press, Boston, 1958 (repr. 1969), p. 5.
3 Catherine M. Howett, 'Notes Towards an Iconography of Regional Landscape Form: The Southern Model', *Landscape Journal*, 4, 2 (1985), p. 75.
4 Kay Anderson and Fay Gale (eds), *Inventing Places*, Longman Cheshire, Melbourne, 1992, Introduction, p. 9.
5 Peter Porter, 'On First Looking into Chapman's Hesiod', in R. Gray and G. Lehmann (eds), *Australian Poetry in the Twentieth Century*, Heinemann, Melbourne, 1991, p. 242.

6 Patrick Dodson, cited in Julian Burger, *The Gaia Atlas of First Peoples*, Penguin, Melbourne, 1990, p. 20.
7 Fustel De Coulanges, *The Ancient City*, Doubleday, New York, 1955, p. 62.
8 Patrick O'Farrell, 'Defining Place and Home. Are the Irish Prisoners of Place?', in David Fitzpatrick (ed.), *Home and Away Immigrants in Colonial Australia*, Australian National University, Canberra, 1992, pp. 13–14.
9 Paul Carter, *The Road to Botany Bay*, Knopf, New York, 1988, p. 154.
10 Mena Abdullah & Ray Mathew, *The Time of the Peacock*, Angus & Robertson, Sydney, 1965, p. 20.
11 Carter, *The Road to Botany Bay*, pp. xxiv, 152–6.
12 J.D. Lang, 'Sonnet Written on Board the Medway off Hobart Town', in R.D. Jordan & P. Pierce (eds), *The Poets' Discovery: Nineteenth Century Australia in Verse*, Melbourne University Press, Melbourne, 1990, p. 153.
13 Mary Fullerton, 'Clearing the Flats', in Patrick Morgan (ed.), *Shadow and Shine, An Anthology of Gippsland Literature*, Centre for Gippsland Studies, 1988, p. 34.
14 Richard Flanagan, 'Wilderness and History', *Public History Review*, I (1992), pp. 111–12.
15 John MacArthur, 'Poem Written on the Banks of the Nepean River, Camden, Australia, September 1827', in Alan Atkinson, *Camden*, Oxford University Press, Melbourne, 1988, p. 25.
16 Atkinson, *Camden*, p. 12.
17 Elizabeth Jolley, 'A Sort of Gift: Images of Perth', in Drusilla Modjeska (ed.), *Inner Cities, Australian Women's Memory of Place*, Penguin, Melbourne, 1989, p. 202.
18 Lucy Turner, cited in Peter Read, 'Being is a Transitive Verb: Four Views of a Life Site in Mittagong New South Wales', *Landscape Research* 19, 2 (Summer 1994), p. 66.
19 Carl Jung, *Memories, Dreams, Reflections*, rec. and ed. Aniela Jaffe, tr. R. & C. Kingston, Collins Fontana, London, 1967, pp. 256–7.
20 Rita Huggins & Jackie Huggins, *Auntie Rita*, Aboriginal Studies Press, Canberra, 1994, p. 13.
21 David Campbell, 'Hawk and Hill', quoted by Joy Hooton, 'Towards an Appreciation of David Campbell's Poetry', in H. Heseltine (ed.), *A Tribute to David Campbell*, University of NSW Press, Sydney, 1987, p. 53.
22 Kate Llewellyn, *The Waterlily*, Hudson, Melbourne, 1987, p. 53.
23 Clare Milner, interview; see also P. Read, 'The Mystery that Seeps from the Earth', paper given at conference (forthcoming).
24 Rex Ingamells, 'Unknown Land', in Brian Elliott (ed.),*The Jindyworobaks*, University of Queensland Press, Brisbane, 1979, p. 21.

25 David Tacey, 'Australian Landscape as a Spiritual Problem', *Canberra Jung Society Newsletter*, Feb.–July 1993, p. 14.
26 Judith Wright, 'The Broken Links', quoted in Peter Read, 'Joy and Forgiveness in a Haunted Country', *New Norcia Studies*, 1994, p. 1.
27 Jill Ker Conway, *The Road to Coorain*, Heinemann, London, 1989, p. 170.
28 'Gurra's Lament', in Ted Strehlow, *Aranda Traditions*, Melbourne University Press, Melbourne, 1947, p. 31.
29 I.P. Halket, *The Quarter Acre Block*, Australian Institute of Urban Studies, Adelaide, 1976, p. 77.
30 Janice Monk, 'Gender in the Landscape; Expressions of Power and Meaning', in Gale and Anderson, *Inventing Places*, p. 123.
31 For example, Gillian Rose, *Feminism and Geography*, University of Minnesota Press, Minneapolis, 1993.
32 Lyn Richards, *Nobody's Home. Dreams and Realities in a New Suburb*, Oxford University Press, Melbourne, 1990, pp. 130–3.
33 Peter Saunders, 'The Meaning of "Home" in Contemporary English Culture', *Housing Studies*, 4, 3 (July 1989), pp. 182–3.
34 Jeanne Kay, 'Landscapes of Women and Men: Rethinking the Regional Historical Geography of the United States and Canada', *Journal of Historical Geography*, 17, 4 (1991), p. 446.
35 Miles Franklin, *My Brilliant Career*, Angus & Robertson, Sydney, 1901 (repr. 1966), p. 164.
36 Katharine Susannah Prichard, *Coonardoo*, Angus & Robertson, Sydney, 1927 (repr. 1975), pp. 52, 100, 127.
37 Maurice Beresford, *The Lost Villages of England*, Lutterworth, London, 1954, p. 64.
38 ibid., pp. 74–5.
39 Gillian Clarke, 'Clwedog', in Tony Curtis (ed.), *The Poetry of Snowdonia*, Bridgend, 1989.
40 Jim Wayne Miller, 'Small Farms Disappearing in Tennessee', in J.W. Miller, *Brier, His Book*, Gnomon Press, 1983, pp. 31–3.
41 Barbara Davison, Hal Kendig, Faye Stephens & Vance Merrill, *It's My Place*, Australian Government Publishing Service, Canberra, 1993, p. 65.
42 For example, see 'Burragorang Defence League, 1941', and other papers held by the Wollondilly Heritage Centre, The Oaks, NSW.
43 Owen W. Pearce, *Rabbit Hot, Rabbit Cold. Chronicles of a Vanishing Australian Community*, Woden, 1991, p. 151.
44 Claude Lee, 'Burragorang Valley', in *A Place to Remember: Burragorang Valley 1957*, Bowral, 1971, pp. 27–8.
45 Pearce, *Rabbit Hot, Rabbit Cold*, p. 142.
46 F. Drake, 'Pomfret Castle', quoted in Anne Janovits, *England's Ruins. Poetic Purpose and the National Landscape*, Blackwell, London, 1990, p. 57.

47 Marc Fried, 'Grieving for a Lost Home', in L.J. Duhl (ed.), *The Urban Condition. People and Policy in the Metropolis*, Basic Books, New York, 1963, pp. 151–2, 167.
48 Jim Hall, in *Valley People*, Kangaroo Press, Kenthurst, 1984, p. 141.
49 Rita & Jackie Huggins, *Auntie Rita*, pp. 156–7.
50 Laurence Brunero, 'A Valley's Epitaph', in Pearce, *Rabbit Hot, Rabbit Cold*, p. 139.
51 Robert Dessaix, in G. Papaellinas (ed.), *Homeland*, Allen & Unwin, Sydney, 1991, p. 153.
52 John A. Agnew, 'The Devaluation of Place in Social Science', in John A. Agnew & James A. Duncan, *The Power of Place*, Unwin Hyman, Boston, 1989, pp. 13–19.
53 Edward M. Soja, *Postmodern Geographies: The Reassertion of Space in Contemporary Social Theory*, Verso, London, 1989, p. 11.
54 J. Grant, 'Gippsland Revisited', in Graeme Kinross Smith & Jamie Grant, *Turn Left at Any Time with Care*, University of Queensland Press, Brisbane, 1975, p. 71.

2 VANISHED HOMELANDS

1 The narrative of Luka Prkan is drawn from interviews and discussions with him. At his request, the name is fictitious.
2 Adapted from Luka Prkan, interview, and Stephanie Thompson, *Australia through Italian Eyes*, Oxford University Press, Melbourne, 1980, pp. 31ff.
3 Jerzy Zubrzycki, *Settlers of the La Trobe Valley*, quoted by Geoffrey Sherington, *Australia's Immigrants*, Allen & Unwin, Sydney, 1980, pp. 147–8.
4 R.A. Baggio, *The Shoe in My Cheese*, R.A. Baggio, Footscray, 1989, pp. 17–19, 87–96.
5 Charles Price (ed.), *Australian Immigration*, Australian National University Press, Canberra, 1966, pp. A5–A7.
6 Phillip Kunz, 'Immigrants and Socialisation: A New Look', *Sociological Review*, 16, 3 (November 1968), p. 363.
7 E.P. Hutchinson, quoted in Kunz, 'Immigrants and Socialisation', p. 368.
8 Andrew Riemer, *Inside Outside*, Angus & Robertson, Sydney, 1992, p. 32.
9 Jean Martin, *Refugee Settlers*, Australian National University Press, Canberra, 1965, p. 5.
10 ibid., pp. 90–1.
11 Nancy Viviani, *The Vietnamese in Australia: New Problems in Old Forms*, Griffith University, Nathan, 1980, p. 13.
12 Martin, *Refugee Settlers*, p. 102.
13 Naomi Rosh White, *From Darkness to Light*, Collins Dove, Melbourne, 1988, pp. 21, 191, 216.

14 ibid., p. 196.
15 The narrative of these events is drawn from interviews with Susan Karas, Theodore Karas and Sylvia Deutsch.
16 Two of three survivors of the Czech ghetto-town of Terezin affirmed, during a talk at the Australian War Memorial in 1994, that they had no desire to return to their birth country despite the overthrow of communist rule.
17 Drawn from interviews with Irena Petrovna Sinelnikova and Irina Somina.
18 Jerzy Krupinski & Graham Burrows, *The Price of Freedom: Young Indochinese Refugees in Australia*, Pergamon, Sydney, 1986, pp. ix, 11. In March 1982 the Australian and Vietnamese governments signed an agreement to allow relatives of Australian Vietnamese to migrate to Australia.
19 The transcripts of these interviews with Vietnamese refugees are held in Melbourne's Living Museum of the West.
20 Lesleyanne Hawthorne (ed.), *Refugee, the Vietnamese Experience*, Oxford University Press, Melbourne, 1982, pp. 316–17.
21 Marysia, in Rosh White, *From Darkness to Light*, p. 193.
22 Judah Waten, *Alien Son*, Angus & Robertson, Sydney, 1952, p. 37.
23 Mena Abdullah & Ray Mathew, *The Time of the Peacock*, Angus & Robertson, Sydney, 1965, p. 90.
24 Jill Matthews, *Good and Mad Women*, Allen & Unwin, Sydney, 1984, pp. 40–3.
25 Drawn from interviews with Cam Trang Dinh and her son Thanh.
26 Martin, *Refugee Settlers*, p. 101.
27 Ruth Johnston, *Immigrant Assimilation*, Peterson Brokensha, Perth, 1965, pp. 24, 137.
28 Harvey Cox, 'The Restoration of a Sense of Place', *Ekistics* 25 (1968), p. 423.
29 Rosh White, *From Darkness to Light*, pp. 159, 46.
30 Interview, Do Thi Anh.
31 This perception has been gained from the responses of Vietnamese interviewees, held at the Living Museum of the West, Melbourne; yet a survey of forty-eight searching questions directed at Indo-Chinese refugees arriving in Darwin in 1980 ethnocentrically omitted the question 'Would you like one day to return to your homeland?'
32 Interview, Bruce Clayton-Brown.
33 For examples, see Coral Edwards & Peter Read (eds), *The Lost Children*, Doubleday, Sydney, 1989.
34 Barbara Schenkel, in Manfred Jurgensen (ed.), *Ethnic Australia*, Phoenix, Brisbane, 1985, p. 102.
35 Waten, *Alien Son*, p. 88.
36 Cornelius Vleeskens, 'This Eternal Curiosity: The Search for a Voice in the

Wilderness', in S. Gunew & Kateryna Longley (eds), *Striking Chords*, Allen & Unwin, Sydney, 1991, pp. 191, 195.

37 Eva Isaacs, *Greek Children in Sydney*, Australian National University Press, Canberra, 1976, pp. 6–7.

38 Ruth Johnston, *The Assimilation Myth*, Publications of the Research Group for European Migration Problems XIV, Nijhoh, The Hague, 1969, p. 64.

39 This section is drawn from interviews and discussions with Senia Paseta.

40 This section is drawn from interviews and discussions with Con Boekel, and from an interview with Peter and Betty Boekel.

41 Quoted by Rosh White, *From Darkness to Light*, pp. 8–9.

42 Elizabeth Wynnhausen, *Manly Girls*, Penguin, Melbourne, 1989, pp. 43–4.

43 Arnold Zable, *Jewels and Ashes*, Scribe, Newham, 1991, pp. 23–4.

44 ibid., p. 102.

45 ibid., p. 163.

46 Arnold Zable, 'Voices from the Silence', in Cassandra Pybus (ed.), *Columbus' Blindness*, University of Queensland Press, Brisbane, 1994, p. 39.

47 Zable, *Jewels and Ashes*, pp. 18, 186.

48 ibid., p. 74.

49 Andrew Riemer, *The Habsburg Cafe*, Angus & Robertson, Sydney, 1993, p. 18.

50 ibid., pp. 206–13, 251.

51 ibid., pp. 146–7.

3 NAMADGI: SHARING THE HIGH COUNTRY

1 The narrative of Granville Crawford is drawn from interviews and discussions with Granville and Rae Crawford.

2 Klaus Hueneke, *Huts of the High Country*, Australian National University Press, Canberra, 1982, pp. xii–xiv, 224–7.

3 D.J. Mulvaney, *NPA* [National Parks Association] *Bulletin*, 25, 4 (June 1988), p. 6.

4 Quoted by Sue Hodges, 'A Sense of Place', in Don Garden (ed.), *Created Landscapes*, History Institute of Victoria, Melbourne, 1992, p. 73.

5 Bill Hicks, quoted in Bryan Jameson, *Movement at the Station*, Collins, Sydney, 1987, p. 127.

6 ibid., p. 127.

7 John Hepworth, 'The Dancing Valley of the Mountain Horsemen', ibid., p. 137.

8 Dick Johnson, *The Alps at the Crossroads*, Victorian National Parks Association, Melbourne, 1974, pp. 141–3.

9 C. Tighe, 'The Origins of Namadgi National Park', *NPA Bulletin*, 29, 1 (March 1992), pp. 14–21.

10 Bob Brown in D.J. Mulvaney (ed.), *The Humanities and the Australian Environment*, Australian Academy of the Humanities, Canberra, 1991, p. 14.
11 Hodges, 'A Sense of Place', p. 84.
12 K. Frawley, 'The Gudgenby Property and Grazing in the National Park', *NPA Bulletin*, 25, 3 (March 1988), p. 4.
13 N. Esau, *NPA Bulletin*, 25, 4 (June 1988), p. 7.
14 D.J. Mulvaney, *NPA Bulletin*, 25, 4 (June 1988), p. 6.
15 R. Story, *NPA Bulletin*, 25, 5 (September 1988).
16 ACT Parks and Conservation Service, 'Management Plan for Namadgi National Park', photocopied typescript, 1986, p. 37.
17 *NPA Bulletin*, 29, 1 (March 1992); see also Tom Griffiths, 'History and National History: Conservation Movements in Conflict', in J. Rickard & P. Spearritt (eds), *Packaging the Past*, Melbourne University Press, Melbourne, 1991, pp. 16–32.
18 ACT Parks and Conservation Service, 'Management Plan', p. 47.
19 National Capital Development Commission, 'The Gudgenby Area. Policy Plan and Development Plan', September 1984, p. 19.
20 *Sydney Morning Herald*, 24 June 1993.
21 Margules & Deverson Pty Ltd, 'Proposed Gudgenby National Park Land Use Study', Australian National Parks and Wildlife Service, Canberra, 1976, p. 15.
22 National Capital Development Commission, 'The Gudgenby Area.' p. 22.
23 Klaus Hueneke, *Kiandra to Kosciusko*, Tabletop Press, O'Connor, 1987, p. 33.
24 'Proposed Gudgenby National Park Land Use Study', pp. 6, 48.
25 Theodore Roosevelt, quoted in Johnson, *The Alps at the Crossroads*, p. 14.
26 Tom Griffiths, 'History and Natural History: Conservation Movements in Conflict?', in Rickard & Spearritt (eds), *Packaging the Past?*, pp. 26–7.
27 ibid., p. 27.
28 Ernie Constance, 'Where the Snowy Mountains Rise', in *When the Currawongs Come Down*, Burrunga Records.
29 'Namadji, Canberra's Highland Wilderness', *Sydney Morning Herald*, 24 June 1993; Granville Crawford's unpublished reply, 6 July 1993.
30 Ms in the possession of Granville Crawford.
31 Johann Kamminga, 'Aboriginal Settlement and Prehistory of the Snowy Mountains', in Babette Scougall (ed.), *Cultural Heritage of the Australian Alps*, Australian Alps Liaison Committee, Canberra, 1991, pp. 106–7.
32 M. Pearson, Seen Through Different Eyes, PhD thesis, Australian National University, Canberra, 1981, p. 87.
33 Josephine Flood, 'Aboriginal Cultural Heritage of the Australian Alps: An Overview', in Scougall, *Cultural Heritage*, p. 85.

34 Fire-stick farming, or periodic burning, gradually reduced forest cover and replaced it with more open grassland..
35 Ken Taylor, 'Cultural Values in Natural Areas', in Scougall, *Cultural Heritage*, pp. 62–3.
36 Kamminga, 'Aboriginal Settlement', pp. 103–14.
37 Richard Helms 1890, 1895, quoted in Josephine Flood, *The Moth Hunters*, Australian Institute of Aboriginal Studies, Canberra, 1980, p. 68.
38 Ann Jackson-Nakano, The Death and Resurrection of the Ngunnawal: A Living History, M Litt thesis, Australian National University, Canberra, 1994, p. 6.
39 ibid., map 2, p. 30.
40 D.B. Rose, *Dingo Makes Us Human*, Cambridge University Press, Cambridge, 1992, p. 111.
41 Song 53 of the Djanggawul Song Cycle, in R.M. Berndt, *Djanggawul*, Cheshire, Melbourne, 1952, p. 128.
42 ACT Parks and Conservation Service, 'Yankee Hat Walking Track' (brochure 1991); Flood, *The Moth Hunters*, pp. 133–6, 330–3.
43 Flood, *The Moth Hunters*, pp. 149–152. Flood notes that the site was 'almost certain to have been a ceremonial ground, and its remoteness makes it seem probable that it was used for initiation purposes'.
44 Rose, *Dingo Makes Us Human*, pp. 106–7.
45 D.J. Mulvaney, 'The Alpine Heritage in Perspective', in Scougall, *Cultural Heritage*, p. 9.
46 Lovell, quoted in Hodges 'A Sense of Place', p. 78.
47 Jameson, *Movement at the Station*, p. 126.
48 Johnson, *The Alps at the Crossroads*, p. 12.
49 For a discussion of the sense of belonging of other Australian farmers, see Peter Read, 'Being is a Transitive Verb', *Landscape Research*, 19, 2 (Summer 1994), p. 61.
50 Thoreau, Journal, 21 July 1851, in Henry Canby (ed.), *The Works of Thoreau*, Houghton Mifflin, Boston, 1937, p. 598.
51 William Wordsworth, 'The Prelude', IV, II, 388–93, quoted in Christopher Salvesen, *The Landscape of Memory*, Edward Arnold, London, 1965, p. 66.
52 Ms, read by Granville Crawford to the author.
53 Interviews with Albert Mullett and Colin Mullett.
54 'The Bimberi Wilderness', Namadgi Visitors Centre.
55 *Imagining Namadgi* (film).
56 'A Mountain Selection', Namadgi Visitors Centre.
57 ACT Parks and Conservation Service, 'Your Guide to the Bush Treasures of Namadgi' (brochure), 1995.

4 TWO DEAD TOWNS

1 The information on Adaminaby, unless otherwise noted, is drawn from interviews with Nita Stewart, Paddy Kerrigan, Norman and Jean Ware, Jock and Pat Wilson, Neville Locker, Merle Russell, Seamus O'Kane, Bill Brooks, Mr and Mrs Leo Crowe, Roy and Elizabeth Ecclestone, Margaret Unger, Leo Stewart, Jan Lucas, Marge and Keith Mackay, Ken Prendergast, Bert and Lillian Tozer and Ron and Trixie Clugston. See also Peter Read, 'Our Lost Drowned Town in the Valley', *Public History Review*, 1 (1992), pp. 160–73.

2 Interviews with Nita Stewart, Paddy Kerrigan, Merle Russell, Mr and Mrs Leo Crowe, Bert and Lillian Tozer, Marge and Keith Mackay.

3 The information on Yallourn is drawn, unless otherwise noted, from materials held by the Centre for Gippsland Studies (CGS), Monash University, Churchill Campus, Vic. The information held there included, in 1994, the exhibition 'Yallourn Revisited'; see also State Electricity Commission of Victoria, 'Report on the Establishment of a Township at Yallourn', 1921, pp. 3–4.

4 Anne Hollensen, 'The Recollections of Jock Lawson', photocopy, 1988, Cranney Collection.

5 ibid., see also Prue McGoldrick, *Yallourn Was*, Gippsland Printers, Morwell, 1984, p. 23.

6 ABC radio talkback with Penny Johnson, the author, and former Yallourn residents, 30 March 1994.

7 Climbing record in Tom Shaw, 'Natural Comedians', short story, typescript, in CGS Archives.

8 Hilary Shaw, interview transcribed in 'Yallourn Revisited' (exhibition).

9 For instance, evidence of Ada Heale and Margaret Longmore in CGS Archives.

10 'Report of the Royal Commission into the Place of Origin and the Causes of the Fires which commenced at Yallourn ...,' 1944, p. 3.

11 Head gardener, in Hollensen, 'The Recollections'; Bert Tayor, interview transcript held in CGS Archives; Councillor Reg Lord, in *Live Wire* (Yallourn newspaper), 30 November 1967.

12 Interviews, Mr and Mrs Leo Crowe; Sioban McHugh, *The Snowy, the People Behind the Power*, Heinemann, Sydney, 1989, p. 219.

13 Bert Taylor and Ken Murray, transcripts of interviews, held at CGS Archives.

14 *Live Wire*, 11 October, 8 November 1961.

15 Stephen Keller, 'Prelude to Power', ABC radio documentary, c. 1959, held in Snowy Mountains Corporation Library, Cooma.

16 The officials did not publish their memoirs of these times until much later, or not at all. For example, the brief recollections of the consulting

architect Don Maclurcan were not published until 1989, in Margaret Unger, *Voices from the Snowy*, University of New South Wales Press, Sydney, 1989, p. 121.

17 Quoted by McHugh, *The Snowy*, p. 212.
18 Soundtrack of film *Operation Adaminaby*, 1958 Snowy Mountains Authority film production, held at Snowy Mountains Corporation Library.
19 D.H. White, *Operation Adaminaby*.
20 D.H. White, draft and amendments, 'The Town of Adaminaby and the Snowy Mountains Scheme', typescript in Snowy Mountains Corporation Library, Cooma, p. 3.
21 Ken Murray, CGS Archives; SEC statement on the future of Yallourn, photocopy, c. September 1970, Cranney Collection.
22 Prue McGoldrick, in Save Yallourn Committee, *To Yallourn with Love*, Mitchell River Press, Bairnsdale, 1984, p. 116.
23 Quoted in Pat Cranney, 'The Yallourn Story', typescript of play, 1989, p. 86, CGS Archives.
24 A.D. Spaull, 'Statements to the Parliamentary Works Committee ...', in 'Slab' (Ted Hopkins), *The Yallourn Stories*, Champion Books, Melbourne, c. 1982, Appendix.
25 'Yallourn Coal Reserves Inquiry', transcript of evidence given before the Public Works Committee at Yallourn, 16 September, 21 October, 4, 9, 25 November 1970, CGS, Cranney Collection.
26 'Bewildering', written statement prepared for the author by Elizabeth Ecclestone; interview, Nita Stewart; Hudson in *Cooma Monaro Express*, 8 May 1956.
27 *Cooma Monaro Express*, 26 July 1956.
28 Nita Stewart to author, 9 March 1995.
29 Identified by Nita Stewart, 18 April 1994.
30 Interviews, Nita Stewart, Geoff Yen, Roy and Elizabeth Ecclestone.
31 Don Maclucan, quoted in Unger, *Voices from the Snowy*, p. 121.
32 *Cooma Monaro Express*, 15 November 1957.
33 Phone conversation with Bob Raymond; Robert Raymond, *A Town to be Drowned*, 1957, copy in Snowy Mountains Corporation library.
34 Anonymous speaker, in *A Town Born to Die*, transcript of SEC film, folder, ms 4122, CGS; film produced by SEC, produced and directed by Mary Wilton.
35 *La Trobe Valley Express*, 1 July 1970.
36 SEC statement, c. September 1970, Cranney Collection, CGS.
37 Central Gippsland TLC, 'Report to VTHC on the Yallourn Township', 9 May 1974, Cranney collection, CGS.
38 Form and invitation, 30 September 1974, Cranney Collection, CGS.

39 *La Trobe Valley Express*, 29 August 1974.
40 Interview, Joe and Pat Dell; songs reproduced in Cranney, 'The Yallourn Story', p. 162, CGS.
41 Interview, Joe and Pat Dell, CGS.
42 *La Trobe Valley Express*, 6 August, 20 August, 5 September 1975; Report by Independent Arbitrator Ray Burkitt, 'Replacement of Yallourn Community Facilities in Nearby Localities', 1 August 1976, copy in CGS Archives.
43 Interview, Bernadette Inglis, née McLaughlin.
44 Interview, Alan Lucas.
45 Interview, David Andrew.
46 Interviews, residents of Adaminaby, including Roy and Elizabeth Ecclestone.
47 'Jack Bridle's Farewell to Talbingo', in *The Settlers Sing More Songs of the Snowy Mountains*, RCA, CAM 128.
48 Paul Connerton, *How Societies Remember*, Cambridge University Press, Cambridge, 1989, p. 15.
49 Nita Stewart to author, 9 March 1995.
50 Nita Stewart, undated published letter, *Cooma Monaro Express*, February 1995.
51 Sally Roberts Jones, 'Tryweryn', in Tony Curtis (ed.), *The Poetry of Snowdonia*, Bridgend, 1989 (hiraeth = sentiment).
52 The residents had been presented with an apparent fait accompli (which in fact had no legal basis) that the town would be 'moved': their responsibility was not to question the scheme but to choose a site: Snowy Mountains Authority, *The Snowy Mountains Story*, SMA, Cooma, 1970s, p. 16.
53 'Chifley', see Country Women's Association, 'The Story of Country Women in Adaminaby 1931–1981', Cooma, 1981.
54 Interview, Paddy Kerrigan.
55 Interview, Nita Stewart.
56 M. Mackay, 'The Old Town', read to Peter Read by the poet.
57 'Slab', *The Yallourn Stories.*
58 From an untitled poem by Urzula Horbach, in Save Yallourn Committee, *To Yallourn With Love*, p. 116.
59 Both events are recounted in 'Slab', *The Yallourn Stories.* The psychic spent the rest of the day recovering by lying on the lawn with a briquette on his chest.
60 *Born To Die.*
61 'The Yallourn Story', p. 70, CGS archives; interview, Pat Cranney; video of opening night held by CGS.
62 'The Yallourn Story', p. 163.
63 David Wardley & Michael Ballock, 'Satisfaction and Positive Resettlement:

Evidence from Yallourn, Latrobe Valley, Australia', *American Planning Association Journal*, 46, 1 (January 1980), pp. 64–75.

64 Prue McGoldrick to author, 13 April 1994.

5 HOME: THE HEART OF THE MATTER

1 Anna Couani, 'Parramatta Sestina', in Drusilla Modjeska (ed.), *Inner Cities, Australian Women's Memory of Place*, Penguin, Melbourne, 1989, p. 16.

2 Drusilla Modjeska, 'Living On a Corner', in Modjeska, *Inner Cities*, p. 67.

3 Mrs McIndoe, interviewed 20 September 1984 for Melbourne's Living Museum of the West, Archives.

4 Quoted in Barbara Davison, Hal Kendig, Faye Stephens & Vance Merrill, *It's My Place*, Australian Government Publishing Service, Canberra, 1993, p. 52.

5 Irene Bright, interviewed 16 July 1984 for Melbourne's Living Museum of the West, Archives.

6 Interview, Trish Gillard.

7 Interview, Peter and Sue Boekel.

8 Interview, Paul and Eleanor Gray.

9 Marie J. Pitt, 'The Hill', in Marie J.Pitt, *Selected Poems of Marie J. Pitt*, Lothian Publishing, Melbourne, 1944, p. 86.

10 Dorothy Hewett, 'In Summer', in Dorothy Hewett, *A Tremendous World in her Head*, Dangaroo Press, Sydney, 1989, pp. 94–5.

11 Judith Wright, 'Old House', in Judith Wright, *A Human Pattern*, Angus & Robertson, Sydney, 1990, pp. 49–50.

12 Nan McDonald, 'The Haunted House', in Nan McDonald, *Pacific Sea*, Angus & Robertson, Sydney, 1947, p. 60.

13 Geoffrey Dutton, 'Anlaby', in Geoffrey Dutton, *Antipodes in Shoes*, Edwards & Shaw, Sydney, 1958, pp. 81–2.

14 Marjorie Pizer, 'On Revisiting My Childhood Home After Many Years', in Marjorie Pizer, *Selected Poems 1963–1983*, Pinchgut Press, Sydney, 1984, p. 45.

15 Philip Hodgins, 'Going Back for a Look', in Philip Hodgins, *Down the Lake with Half a Chook*, ABC Enterprises, Sydney, 1988, p. 69.

16 Edward Relph, 'Geographical Experiences and Being-in-the World: The Phenomenological Origins of Geography', in David Seamon & Robert Mugerauer (eds), *Dwelling, Place and Environment*, Martinus Nijhoff Publishers, Dordrecht, 1985, pp. 26–7.

17 This section is drawn from Anne Boyd, interview and visit.

18 J. Douglas Porteous, 'Home the Territorial Core', *Geographical Review*, LXVI (1976), pp. 383–90.

19 Interviews, Alice and John Rawe.

20 THE ASH WEDNESDAY MEMORIAL PARK.
is dedicated to the determination and courage
of the Macedon community
after the devastating fires
of 16th February 1983

21 Albert Hart, in Chris Healey (ed.), *The Lifeblood of Footscray: Working Lives at the Angliss Meatworks*, Melbourne's Living Museum of the West, Footscray, c. 1982, pp. 137–8.
22 George Linnard, in ibid., pp. 153–4.
23 Unless otherwise noted, the information on Cribb Island is drawn from interviews, Pat and John Jackson.
24 Letter in Mr and Mrs Jackson's possession.
25 'Pat Jackson, 'Cribby', 1994, ms.
26 The information in this section is drawn from Libby Plumley and Barbara Tong, interviews. The spelling was altered to 'Weetangera' by the National Capital Development Commission.
27 Interview, Libby Plumley.
28 Yi-Fu Tuan, *Space and Place*, University of Minnesota Press, Minneapolis, 1979, p. 144.
29 Interview and visit, Prue and Val McGoldrick, 19 April 1994.
30 Doreen Massey, 'A Place called Home', in *New Formations* 17 (Summer 1992), pp. 3–15.
31 J. Nicholas Entrekin, *The Betweeness of Place*, Johns Hopkins University Press, Baltimore, 1991, p. 45.

6 THE INUNDATION OF LAKE PEDDER

1 For a brief history of the campaign to save Lake Pedder, see Burton Committee, *The Future of Lake Pedder*, Report of the Lake Pedder Committee of Inquiry, June 1973, Lake Pedder Action Committees, Hobart, 1973.
2 Tim Bonyhady, 'Lake Pedder 1971', *Island*, 56 (Spring 1993), p. 34.
3 Witness Geoffrey Parr, Burton Committee, Lake Pedder Committee of Inquiry, transcript of evidence, 2–6 April 1973, pp. 46–50.
4 Most of the language being discussed in this section has been drawn from the Burton Committee transcript. Other sources are *Lake Pedder* (film), Tasfilm 1971; Bonyhady, 'Lake Pedder 1971'; Tim Bonyhady, *Images in Opposition, Australian Landscape Painting 1801–1890*, Oxford University Press, Melbourne, 1985; Bob Brown, *Lake Pedder*, The Wilderness Society, Hobart, 1985; Lake Pedder Action Committees of Tasmania and Victoria, pr. and dir. by Peter Dodds and Ross Matthews, *The Struggle for Pedder* (film) 1971; H. Brinsmead-Hungerford, *I Will Not Say the Day is Done*, APCOL, Sydney 1983.

5 *Lake Pedder*, Tasfilm, 1971.
6 Witness Beverley Dunn, in Burton Committee, *The Future of Lake Pedder*, pp. 174–9.
7 Kevin Kiernan, in Brown, *Lake Pedder*, p. 23.
8 Quoted by Malcolm Leggett, Fake Pedder: Drowning or Enlarging a Lake, BA (Hons) thesis, ANU, 1994, p. 40.
9 Leggett quoting Bonyhady, *Images in Opposition*, p. 41.
10 Walter Bruegemann, *The Land: Place as Gift, Promise and Exchange in Biblical Faith*, Fortress Press, Philadelphia, c. 1977, pp. 29, 43.
11 Roderick Nash, *Wilderness and the American Mind*, Yale University Press, New Haven, pp. 10–40.
12 Discussion of Gainsborough's 'Landscape with a Woodcutter Courting a Milkmaid', in John Barrell, *The Idea of Landscape and the Sense of Place*, Cambridge University Press, Cambridge, 1972, pp. 50–3.
13 John Barrell, '"The Public Prospect and the Public View": The Politics of Taste in Eighteenth Century Britain', in Christopher Eade (ed.), *Projecting the Landscape*, Humanities Research Centre, Canberra, 1987, pp. 15–35.
14 Barrell, *The Idea of Landscape*, pp. 57–9.
15 Jay Appleton, *The Experience of Landscape*, John Wiley, London, 1975, pp. 33–40.
16 Karl Kroeber, *Romantic Landscape Vision, Constable and Wordsworth*, University of Wisconsin Press, Madison, 1975, pp. 6–7.
17 Byron, 'Manfred', quoted in Nash, *Wilderness*, p. 47.
18 Raymond Williams, *The Country and the City*, Oxford University Press, Oxford, 1973, pp. 127–30.
19 See ch. 1.
20 Oliver Goldsmith, 'The Deserted Village', II. 395–9, in Tom Davis (ed.), *Oliver Goldmith, Poems and Plays*, Dent, London, 1975, p. 191.
21 Barrell, *The Idea of Landscape*, pp. 94–5.
22 ibid., p. 73.
23 Williams, *The Country and the City*, p. 120.
24 Quoted by Marjorie Nicolson, *Mountain Gloom and Mountain Glory*, Norton, New Haven, 1967, p. 18.
25 Christopher Salvesen, *The Landscape of Memory*, Edward Arnold, London, 1965, pp. 172–95.
26 Kroeber, *Romantic Landscape Vision*, pp. 57–8.
27 David Lowenthal, 'The Place of the Past in the American Landscape', in D. Lowenthal & M.J. Bowden (eds), *Geographies of the Mind, Essays in Historical Geosophy*, Oxford University Press, New York, 1976, p. 102.
28 Thoreau, *Journal*, 21 July 1851, in Henry Canby (ed.), *The Works of Thoreau*, Houghton Mifflin, Boston, 1937, pp. 599–600.

29 Quoted in Nash, *Wilderness*, pp. 125–6.
30 G. Altmeyer, 'Three Ideas of Nature in Canada, 1893–1914', *Journal of Canadian Studies*, 11, 3 (1976), p. 31.
31 ibid., pp. 32–3.
32 Quoted by Colin Hall, *Wasteland to World Heritage*, Melbourne University Press, Melbourne, 1992, pp. 108–9.
33 Williams, *The Country and the City*, pp. 132–4.
34 Yi-Fu Tuan, *Space and Place*, Edward Arnold, London, 1977, p. 54.
35 Barrell, *The Idea of Landscape*, p. 183.
36 William S. Ramson, 'Wasteland to Wilderness: Changing Perceptions of the European Environment', in D.J. Mulvaney (ed.), *The Humanities and the Australian Environment*, Australian Academy of the Humanities, Canberra 1991, pp. 5–19.
37 David Goodman, 'Gold Fever/Golden Fields: The Language of Agrarianism and the Victorian Gold Rush', *Australian Historical Studies*, 29, 90 (April 1988), pp. 21, 34.
38 David Burn, quoted in S. Martin, *A New Land*, Allen & Unwin, Sydney, 1993, p. 125; for another reaction to Mount Wellington, see 'The Wanderer, Supposed to be Written by a Native of Van Dieman's Land', in R.D. Jordan & P. Pierce, *The Poets' Discovery: Nineteenth Century Australia in Verse*, Melbourne University Press, Melbourne, 1990, p. 177.
39 Rosa Praed, *Lady Bridget in the Never Never*, Pandora, London, 1915 (repr. 1987), pp. 143, 167.
40 Bonyhady, *Images in Opposition*, p. 75.
41 ibid., pp. 122ff.
42 ibid., pp. 136–8.
43 Henry Lawson, 'The Bush Undertaker', in Colin Roderick, *Henry Lawson. Short Stories and Sketches*, Angus & Robertson, Sydney, 1972, vol. 1, pp. 55, 57.
44 Eleanor Dark, *The Timeless Land*, Collins, London, 1947, p. 54.
45 Bruce Chatwin, *The Songlines*, Jonathon Cape, London, 1987, p. 46.
46 D.H. Lawrence & M.L. Skinner, *The Boy in the Bush*, Heinemann, London, 1924 (repr. 1972), p. 239.
47 Praed, *Lady Bridget*, p. 63.
48 Beverley Dunn, in Burton Committee, *The Future of Lake Pedder*, pp. 174–5.
49 Kroeber, *Romantic Landscape Vision*, p. 24.
50 UNESCO, World Heritage Committee, 1984, quoted in Hall *Wasteland to World Heritage*, p. 169.
51 ibid., p. 168.
52 Sally Roberts, in Tony Curtis (ed.), *The Poetry of Snowdonia*, Severn Press, Bridgend, 1989.
53 In *The Settlers Sing More Songs of the Snowy Mountains*, RCA CAM 128.

54 Marge Mackay, 'Adaminaby, the Old Town'.
55 Douglas Stewart, 'Farewell to Jindabyne', in Douglas Stewart *Collected Poems 1936–1967*, Angus & Robertson, Sydney, 1967, p. 28.
56 Stephen Edgar, 'Pedder', in Stephen Edgar, *Ancient Music*, Angus & Robertson, Sydney, 1988, p. 17.
57 Gordon Franklin, 'Let the Franklin Flow', 1981.
58 Bernard Smith, 'On Perceiving the Australian Suburb', in G. Seddon & Mari Davis (eds), *Man and Landscape in Australia, Towards an Ecological Vision*, Australian Government Publishing Service, Canberra, 1976, pp. 294–6. See also Ken Taylor, 'Defining an Australian Sense of Place: Cultural Identity in Landscape and Painting', *CELA 94*, History and Culture Conference Proceedings (Council of Educators in Landscape Architecture), pp. 270–80.

7 DARWIN REBUILT

1 Curly Nixon, interview by Francis Good, Northern Territory Archives Service (NTAS), NTRS 226, TS 654.
2 George Brown, interview by Francis Good, NTAS, NTRS 226, TS 572.
3 'Talking History', *Land Rights News*, 2/35 (April 1995), p. 30.
4 Sally Roberts, interview by Suzanne Saunders (Parry), Oral History Unit, NTAS, NTRS 226, TS 567.
5 Interviews, Peter and Kass Hancock.
6 Richard Creswick, interview by Christine Bond, NTAS, NTRS 226, TS 536.
7 Donald Sanders, interviewed by Ronda Jamieson, NTAS, NTRS 226, TS 503.
8 Keith Cole, *Winds of Fury*, Rigby, Adelaide, 1977, pp. 43–4.
9 Suzanne Parry, personal communication.
10 Ruary Bucknell, interview by Francis Good, NTAS, NTRS 226, TS 599.
11 Interview, Ruary Bucknell.
12 Anon., in *Contact* (Overseas Telecommunications newsletter) 4, 2 (second quarter 1975).
13 Some 10 000 people self-evacuated by road. Aboriginal families from Bagot reserve were allowed to evacuate as complete family units.
14 Unnamed interview subject, in Bill Bunbury (prod. and pres.), 'I Still Don't Like High Winds', program 2, ABC Social History Unit, 1991.
15 Alan Stretton, *The Furious Days*, Collins, Sydney, 1976, p. 110.
16 Quoted in Cole, *The Winds of Fury*, p. 47.
17 Dawn Lawrie, 'The Frustrations of a Civil Population Associated with a Major Reconstruction Project', photocopy, c. 1979.
18 Cole, *The Winds of Fury*, p. 49.
19 Lawrie, 'The Frustrations of a Civil Population'.
20 Thelma Crossby, in *Northern Territory News*, Commemorative Supplement, 'Cyclone Tracy: A Story of Survival', c. 12 December 1994, p. 13.

21 Drawn from Max Dumais, *Final Report of the Darwin Disaster Welfare Committee*, Darwin 1976, pp. 218–47.
22 Quoted in Cole, *The Winds of Fury*, p. 68.
23 Commonwealth Newsletter, No. 11, 7 January 1975.
24 Gordon Parker, 'Psychological Disturbance in Darwin Evacuees Following Cyclone Tracy', *Medical Journal of Australia*, 21 (24 May 1975), pp. 650–2; see also editorial, p. 638.
25 Editorial, ibid., p. 639.
26 Gordon Milne, 'Cyclone Tracy I. Some Consequences of the Evacuation for Adult Victims', *Australian Psychologist*, 12, 1 (March 1977), p. 52.
27 Department of Social Security, *The Experience of Cyclone Tracy*, Australian Government Publishing Service, Canberra, 1981, p. 149.
28 ibid., p. 148.
29 D. Webber, 'Darwin Cyclone: An Exploration of Disaster Behaviour', *Australian Journal of Social Issues*, 11, 1 (1976), p. 61.
30 *Bunji* (newspaper), January 1975, in Bill Day, *Bunji*, Aboriginal Studies Press, Canberra, 1994, pp. 69–70.
31 William Walsh, in Bill Bunbury, *Cyclone Tracy: Picking up the Pieces*, Fremantle Arts Centre Press, Fremantle, 1994, p. 127.
32 Interview, George Brown.
33 Dr Ella Stack, quoted in Cole, *The Winds of Fury*, p. 132.
34 'Cyclone Tracy Darwin 1974' (video), Australian Broadcasting Commission.
35 Interview, Sally Roberts.
36 Joint statement, R. Patterson and T. Uren, n.d., cited in Dumais, *Final Report*, pp. 153–4.
37 Ken Todd, 'Darwin Post-Tracy', *Royal Australian Planning Institute Journal* 17 (August 1979), p. 193.
38 J. Cross, A Survey of Darwin's Social History 1868–1956, BA(Hons) thesis, University of Adelaide, 1959, pp. 113–15.
39 Department of Housing and Construction, 'Darwin Reconstruction Study', photocopy, n.d. [18 January 1975], pp. i, 3.
40 ibid., pp. 1–3.
41 Cities Commission, *Planning Options for Future Darwin*, Australian Government Publishing Service, Canberra 28 January 1975, pp. 3–20.
42 Quoted in *Northern Territory News*, Commemorative Supplement, p. 3.
43 Rex Patterson, Minister for the Department of Northern Territory, Second Reading speech, *Darwin Reconstruction Act*.
44 Uren to Patterson, Cities Commission, in Darwin Planning Guidelines No. 2, Prepared on Behalf of the Department of Northern Territory for the Darwin Reconstruction Commission, 14 March 1975, pp. vii–viii; Report p. 30.

45 Barbara James, personal communication.
46 Gerry Tschirner, *Northern Territory News*, Commemorative Supplement.
47 Denys Green, reported in *Northern Territory News*, Commemmorative Supplement.
48 Quoted in Cole, *The Winds of Fury*, p. 104.
49 Quoted ibid., p. 106.
50 Quoted ibid., p. 108.
51 Draft press release, n.d., c. 1 April 1975, Dr Rex Patterson, Minister for the Department of Northern Territory.
52 'Cyclone Tracy', *Weekend Australian*, 17 December 1994.
53 Lawrie, 'The Frustrations of a Civil Population', p. 5.
54 *The New Darwin*, 1, 1 (26 August 1975); Dumais, *Final Report*, pp. 23–32.
55 Dumais, *Final Report*, pp. 28–9.
56 Lawrie 'The Frustrations of a Civil Population', p. 8.
57 Lawrie, quoted in Cole, *The Winds of Fury*, p. 136.
58 Interview, Peter Hancock.
59 Milne, 'Cyclone Tracy I', p. 52.
60 'Cyclone Tracy Twenty Years Later', *Weekend Australian*, Special Supplement, 24 December 1994.
61 *Northern Territory News*, 12 November 1994.
62 Interview in Bunbury, 'I Still Can't Stand High Winds'.
63 According to Sally Roberts, whose house stood on the site, the girders actually came from three houses and were pushed into their extraordinary shape by bulldozers clearing wreckage away a few days after the cyclone.
64 Bunbury, *Cyclone Tracy*.
65 Eric Johnston, in *Northern Territory News*, 2 July 1994, *Canberra Times*, 13 December 1994.
66 Sherylee Armstrong, in *Northern Territory News*, 12 November 1994.
67 John Loizon, in *Northern Territory News*, 6 July 1994.
68 *Northern Territory News*, 30 November 1994.
69 'Cyclone Tracy: A Story of Survival', *Northern Territory News*, Supplement, c. 1 December 1994.
70 *Northern Territory News*, 15, 17 December 1994. On 21 December Rainbow, having been telephoned by the Committee chair, wrote to apologise to the Anniversary Committee, but remained 'disgusted and ashamed' of the attitude of the businesspeople whose 'main aim' was 'to profit from these proceedings'.
71 Interview, Dr Mickey Dewar.
72 Interview, Dawn Lawrie.
73 I am grateful to Samantha Wells for discussion and information on this topic.
74 *Land Rights News*, April 1995.

8 LOSING A NEIGHBOURHOOD

1 George Drossos, in Studs Terkel, *Division Street: America*, Allen Lane, Penguin Press, London, 1968, p. 108.
2 Barry Byrne in ibid., pp. 248–9.
3 Florence Scala in ibid., p. 32.
4 Alan Gilbert, 'The Roots of Anti-Suburbanism in Australia', in S.L. Goldberg & F.B. Smith (eds), *Australian Cultural History*, Cambridge University Press, Cambridge, 1988, pp. 34, 40–43.
5 D.H. Lawrence, *Kangaroo*, Heinemann, Melbourne 1923 (repr. 1963), p. 6.
6 Quoted by Gilbert, 'The Roots of Anti-Suburbanism', pp. 46–7.
7 C. Wallace-Crabbe, 'Melbourne in 1963', quoted in Bruce Bennett, *Place, Region and Community*, James Cook University, Townsville, 1984, p. 43.
8 ibid.
9 Gilbert, 'The Roots of Anti-Suburbanism', p. 45; see also Bernard Smith, 'On Perceiving the Australian Suburb', in G. Seddon & Mari Davis (eds), *Man and Landscape in Australia, Towards an Ecological Vision*, Australian Government Publishing Service, Canberra, 1976.
10 Kenneth Jackson, *The Crabgrass Frontier*, Oxford University Press, New York, p. 79.
11 Gilbert, 'The Roots of Anti-Suburbanism', p. 33.
12 J.M. Freeland, 'People in Cities', in Amos Rapoport (ed.), *Australia as Human Setting*, Angus & Robertson, Sydney, 1972, pp. 115–16.
13 Herbert Gans, *The Urban Villagers*, The Free Press, New York, 1963, ch. 13.
14 Herbert Gans, *People, Places and Policies*, Columbia University Press, 1991, pp. 210ff.
15 Shirley Fitzgerald, *Sydney 1842–1992*, Hale & Iremonger, Sydney, 1992, pp. 109–20.
16 ibid., p. 224.
17 ibid., p. 222–3.
18 Interview, Julia Martin.
19 Amy Phillips, in Carlton Forest Project, 'Carlton People and Social Change', No. 6, 1988, p. 21.
20 ibid., pp. 7–8, 22.
21 Tony Birch, 'The Battle for Spatial Control in Fitzroy', in Sarah Ferber, Chris Healy & Chris McAuliffe (eds), *Beasts of Suburbia*, Melbourne University Press, Melbourne, 1994, p. 34; the quotation is by former resident John Kyrious.
22 Frontispiece, in Council of the City of Sunshine, *Sunshine Cavalcade* (souvenir booklet commemorating the proclamation of the city of Sunshine) c. 1951.
23 Gans, *People, Places and Policies*, p. 57.

24 Jackson, *The Crabgrass Frontier*, pp. 238–9.
25 M. Crozier, 'The Idea of a Garden', *Meanjin*, 47, 3 (1988), p. 398; D. Goodman, 'The Politics of Horticulture', *Meanjin*, 47, 3 (1988), pp. 403–12.
26 Hugh Stretton, quoted by Jean Duruz, 'Suburban Gardens: Cultural Notes', in Ferber, Healey & McAuliffe, *Beasts of Suburbia*, p. 198.
27 Mrs Rolf Boldrewood, quoted by S. Hosking, 'I 'ad to 'ave me Garden', *Meanjin*, 47, 3 (1988), p. 441.
28 Helen Proudfoot, *Gardens in Bloom*, Kangaroo Press, Sydney, 1989, pp. 1–22.
29 Cf the interesting discussion on gender/power relations in Duruz, 'Suburban Gardens: Cultural Notes'.
30 Walter Murdoch 1921, quoted in Bennett, *Place, Region and Community*, p. 42.
31 Lawrence, *Kangaroo*, pp. 6–7, 20.
32 Advertising dodger, 'Magnificent Mundroola Estate', c. 1964, original in the possession of Noel Snelling.
33 NSW Roads and Traffic Authority, *F2-Castlereagh Freeway, Environmental Impact Statement*, NSW Government Printer, Sydney, 1989, pp. 26–34.
34 Interview, Esme Blackmore; Department of Main Roads to J.A. Blackmore, 15 January 1968.
35 Interview, Ivan and Jenny Lewis.
36 Tom Uren, Minister for Lands, to Ivan Lewis, 18 January 1966.
37 Interview, Julia Garnett.
38 Roads and Traffic Authority, *F2-Castlereagh Freeway*, p. 1.
39 *Environmental Planning and Assessment Act 1979 (NSW)*, s. 112 (1); clause 57 of the regulation; quoted in Roads and Traffic Authority, F2-Castlereagh Freeway, pp. 119–22.
40 Roads and Traffic Authority, *F2-Castlereagh Freeway*, p. 114.
41 Friends of Beecroft Reserve, *Newsletter*, 6 July 1989.
42 John Woodward, Commissioner of Inquiry for Environment and Planning, Report to the Honourable David Hay, *A Proposed Expressway from Pennant Hills Road, Beecroft to Pittwater Road, Ryde, known as the F2 Stage 1*, July 1990, NSW Government Printer, 1990, pp. 1–6, 229.
43 Roads and Traffic Authority, F2-Castlereagh Freeway, pp. 82–3.
44 Personal visit with Julia Garnett, January 1995.
45 Roads and Traffic Authority, F2-Castlereagh Freeway, pp. 54–5.
46 Woodward, *A Proposed Expressway*, pp. 4–5, 96–107.
47 ibid., pp. 230–3.
48 Definition of wilderness in the Namadgi Visitors Centre, ACT; see also, ch. 3.
49 Julia Garnett, 'Bushland at Work', Friends of Beecroft Reserve, photocopy, n.d., in possession of the author.

50 'Careerism, Incompetence or Funny Finance', *Urban Action*, Special Supplement, 8 (Winter 1993), pp. 1–2.
51 J. Goldberg, 'Sydney Motorways cannot be Justified', *Australian Financial Review*, 21 October 1993; 'The F2 Castlereagh Expressway Affair: A Case for Reform of the NSW Roads and Traffic Authority', *Urban Policy and Research*, 11, 3 (1993), pp. 132–49.
52 Roads and Traffic Authority, *Community Newsletter No. 2*, December 1993, p. 3.
53 'Careerism', p. 1.
54 Goldberg, 'The F2 Castlereagh Expressway Affair', p. 146.
55 Julia Garnett, Extract from 'The Turpentine Tree'.
56 Interview, Esme Blackmore.

9 THAT PLACE

1 Kenneth Frampton, 'Towards a Critical Regionalism: Six Points for an Architecture of Resistance', in Hal Foster (ed.), *The Anti-Aesthetic. Essays on Post-modern Culture*, Bay Press, Port Townsend, 1983, pp. 17, 26.
2 For example, Kevin Robins, 'Reimagined Communities? European Image Spaces, Beyond Fordism', *Cultural Studies 3*, 2 (May 1989), pp. 145–65.
3 Frampton, 'Towards a Critical Regionalism', p. 24.
4 D.B. Rose, 'Review of K. Neumann *Not the Way It Really Was*', *Australian Historical Studies*, 103 (October 1994), p. 311.
5 'Avram', in Naomi Rosh White, *From Darkness to Light*, Collins Dove, Melbourne, 1988, p. 190.
6 National Parks and Wildlife Service, 'Proposed Gudgenby National Park Land Use Study', ANPWS, Canberra, 1976, pp. 6, 48.
7 Editorial, *Medical Journal of Australia*, 24 May 1975, p. 639.
8 Roads and Traffic Authority, *Community Newsletter No. 2*, December 1993, p. 3.
9 Gerda M. Speller, personal communication.
10 Gerda M. Speller, Landscape, Place and the Psycho-social Impact of the Channel Tunnel Terminal Project, MSc, University of Surrey, 1988, pp. 1–3, 49.
11 Anne Morrow Lindbergh, *Hour of Gold, Hour of Lead*, Harcourt Brace Jovanovich, New York, 1973, p. 214.
12 Report of the Committee of Inquiry into the National Estate, Australian Government Publishing Service, Canberra, 1974, Rec. 27, 27(g), p. 338.
13 Criteria for the Register of the National Estate, Nos E, F, G, H, repr. in Australian Heritage Commission, *More than Meets the Eye*, 1993 Technical Workshop Series No. 7, p. 87.
14 Alison Rawe, personal communication.

15 News item, *The 7.30 Report*, ABC television, 9 February 1995.
16 Suzanne Parry, personal communication. Presumably a coin collection in the home formerly standing at the site had crashed to the ground at that point.
17 Anne Boyd to author, 16 February 1995.
18 *Sydney Morning Herald*, 27 April 1995; see also *Northern Herald*, 27 April 1995.
19 Julia Garnett, personal communication.
20 Prue McGoldrick to author, 13 April 1994.

REFERENCES

INTERVIEWS AND DISCUSSIONS

David Andrew, Do Thi Anh, Pat Arthur, Esme Blackmore, Con and Trish Boekel, Peter and Sue Boekel, Peter and Betty Boekel, Anne Boyd, Bill Brooks, Roy and Trixie Clugston, Rhonia Craig, Pat Cranney, Granville and Rae Crawford, Mr and Mrs Leo Crowe, Sylvia Deutsch, Mickey Dewar, Cam Trang Dinh, Thang Duc Do, Roy and Elizabeth Ecclestone, Julia Garnett, Trish Gillard, Braham Glass, Paul and Eleanor Gray, Peter and Kass Hancock, Bernadette Inglis, Barbara James, Margaret and Jim Johnson, Susan and Theodore Karas, Paddy Kerrigan, Dawn Lawrie, Ivan and Jenny Lewis, Neville Locker, Alan Lucas, Marge and Keith Mackay, Julia Martin, Prue and Val McGoldrick, Claire Milner, Albert and Colin Mullett, Yung Thin Nguyen, Seamus O'Kane, Suzanne Parry, Senia Paseta, Libby Plumley, Luka Prkan, Ken Prendergast, Alice and John Rawe, Sally Roberts, Phil Robinson, Merle Russell, Irena Petrovna Sinelnikova, Irina Somina, John and Pam Squirrell, Nita Stewart, Barbara Tong, Bert and Lillian Tozer, Milka and Stipa Ujdur, Margaret Unger, Gwen and Geoff Walker, Norman and Jan Ware, Marivic Wyndham, Pat and Jock Wilson, Geoff Yen.

BIBLIOGRAPHY OF WORKS CITED

Abdullah, Mena & Ray Mathew, *The Time of the Peacock*, Angus & Robertson, Sydney, 1965.

ACT Parks and Conservation Service, 'Management Plan for Namadgi National Park', typescript, 1986.

—— 'Yankee Hat Walking Track', brochure, 1991.

—— 'Your Guide to the Bush Treasures of Namadgi', brochure, 1995.

ACT Parks and Wildlife Service, *Imagining Namadgi*, film, c. 1992.

Agnew, John A., 'The Devaluation of Place in Social Science' in John A. Agnew & James A. Duncan (eds), *The Power of Place*, Unwin Hyman, Boston, 1989.

Altmeyer, G., 'Three Ideas of Nature in Canada', 1893–1914', *Journal of Canadian Studies*, 11, 3 (1976), pp. 21–36.

Anderson, Kay & Fay Gale (eds), *Inventing Places*, Longman Cheshire, Melbourne, 1992.

Anon., *Contact* [Overseas Telecommunications newsletter] 4, 2 (second quarter 1975).

Appleton, Jay, *The Experience of Landscape*, John Wiley, London, 1975.

Atkinson, Alan, *Camden*, Oxford University Press, Melbourne, 1988.

Australian Heritage Commission, *More than Meets the Eye*, 1993 Technical Workshop Series No. 7.

Bachelard, Gaston, *The Poetics of Space*, tr. Maria Jolas, Beacon Press, Boston, 1958 (repr. 1969).

Baggio, R.A., *The Shoe in my Cheese*, R.A. Baggio, Melbourne, 1989.

Barrell, John, *The Idea of Landscape and the Sense of Place*, Cambridge University Press, Cambridge, 1972.

——' "The Public Prospect and the Public View": The Politics of Taste in Eighteenth Century Britain', in Christopher Eade (ed.), *Projecting the Landscape*, Humanities Research Centre, Canberra, 1987, pp. 15–35.

Bennett, Bruce, *Place, Region and Community*, James Cook University, Townsville, 1984.

Beresford, Maurice, *The Lost Villages of England*, Lutterworth, London, 1954.

Berndt, R.M., *Djanggawul*, Cheshire, Melbourne,1952.

Birch, Tony, 'The Battle for Spatial Control in Fitzroy' in Sarah Ferber, Chris Healy and Chris McAuliffe (eds), *Beasts of Suburbia*, Melbourne University Press, Melbourne, 1994, pp. 18–34.

Bonyhady, Tim, 'Lake Pedder 1971', *Island*, 56 (Spring 1993).

——*Images in Opposition, Australian Landscape Painting 1801–1890*, Oxford University Press, Melbourne, 1985.

Bridle, Jack, 'Jack Bridle's Farewell to Talbingo', in *The Settlers Sing More Songs of the Snowy Mountains*, RCA, CAM 128.

Brinsmead-Hungerford, H., *I Will Not Say the Day is Done*, APCOL, Sydney, 1983.

Brown, Bob, *Lake Pedder*, The Wilderness Society, Hobart, 1985.

Bruegemann, Walter, *The Land: Place as Gift, Promise and Exchange in Biblical Faith*, Fortress Press, Philadelphia, c. 1977.

Bunbury, Bill, *Cyclone Tracy: Picking Up the Pieces*, Fremantle Arts Centre Press, Fremantle, 1994.

Bunbury, Bill (prod. and pres.), 'I Still Don't Like High Winds', ABC Social History Unit, Perth, 1991.

Burger, Julian, *The Gaia Atlas of First Peoples*, Penguin, Melbourne, 1990.

Burkitt, Ray, *Report by Independent Arbitrator Ray Burkitt, Replacement of Yallourn*

Community Facilities in Nearby Localities, 1 August 1976, photocopy in Centre for Gippsland Studies.

Burton Committee, *The Future of Lake Pedder*, Report of the Lake Pedder Committee of Inquiry, June 1973, Lake Pedder Action Committees, Hobart, 1973.

Canby, Henry (ed.), *The Works of Thoreau*, Houghton Mifflin, Boston, 1937.

Carlton Forest Project, *Carlton People and Social Change*, 6, 1988.

Carter, Paul, *The Road to Botany Bay*, Knopf, New York, 1988.

Centre for Gippsland Studies, 'Yallourn Revisited', exhibition, 1994.

Chatwin, Bruce, *Songlines*, Jonathon Cape, London, 1987.

Chekhov, Anton, 'The Cherry Orchard', in *Three Plays*, Penguin, Harmondsworth, 1951.

Cities Commission, *Planning Options for Future Darwin*, Australian Government Publishing Service, Canberra, 28 January 1975.

Cole, Keith, *Winds of Fury*, Rigby, Adelaide, 1977.

Commonwealth Newsletter [Darwin], 11 (7 January 1975).

Connerton, Paul, *How Societies Remember*, Cambridge University Press, Cambridge, 1989.

Constance, Ernie, 'When the Snowy Mountains Rise', in *When the Currawongs Come Down*, Burrunga Records.

Cooma Monaro Express, 8 May, 26 July 1956, 15 November 1957.

Council of the City of Sunshine, *Sunshine Cavalcade*, Sunshine, c. 1951.

Country Women's Association, Adaminaby, 'The Story of Country Women in Adaminaby 1931–1981', Cooma, 1981.

Cox, Harvey, 'The Restoration of a Sense of Place', *Ekistics*, 25 (1968), pp. 422–4.

Cranney, Pat, 'The Yallourn Story', ms, 1989, Centre for Gippsland Studies.

Cross, J., A Survey of Darwin's Social History 1868–1956, BA (Hons) thesis, University of Adelaide, 1959.

Crozier, M., 'The Idea of a Garden', *Meanjin*, 47, 3 (1988), pp. 397–402.

Curtis, Tony (ed.), *The Poetry of Snowdonia*, Severn Press, Bridgend, 1989.

Dark, Eleanor, *The Timeless Land*, Collins, London, 1947.

Davis, Tom (ed.), *Oliver Goldsmith, Poems and Plays*, Dent, London, 1975.

Davison, Barbara, Hal Kendig, Faye Stephens & Vance Merrill, *It's My Place*, Australian Government Publishing Service, Canberra, 1993.

Day, Bill, *Bunji*, Aboriginal Studies Press, Canberra 1994.

De Coulanges, Fustel, *The Ancient City*, Doubleday, New York, 1955.

Department of Housing and Construction, NT Region, *Report*, 1974–76.

—— 'Darwin Reconstruction Study', photocopy, n.d. [18 January 1975].

Department of Northern Territory for the Darwin Reconstruction Commission, 'Darwin Planning Guidelines', 2, 14 March 1975.

Department of Social Security, *The Experience of Cyclone Tracy*, Australian Government Publishing Service, Canberra, 1981.

Duhl, L.J. (ed.), *The Urban Condition. People and Policy in the Metropolis*, Basic Books, New York, 1963.

Dumais, Max (ed.), *Final Report of the Darwin Disaster Welfare Council*, Darwin, 1976.

Duruz, Jean, 'Suburban Gardens: Cultural Notes', in Sarah Ferber, Chris Healy & Chris McAuliffe (eds), *Beasts of Suburbia*, Melbourne University Press, Melbourne, 1994, pp. 198–213.

Dutton, Geoffrey, *Antipodes in Shoes*, Edwards & Shaw, Sydney, 1958.

Eade, Christopher (ed.), *Projecting the Landscape*, Humanities Research Centre, Canberra, 1987.

Edgar, Stephen, *Ancient Music*, Angus & Robertson, Sydney, 1988.

Edwards, Coral & Peter Read (eds), *The Lost Children*, Doubleday, Sydney, 1989.

Electricity Commission of Victoria, *Report on the Establishment of a Township at Yallourn*, Victorian Government Printer, 1921.

Elliott, Brian (ed.), *The Jindyworobaks*, University of Queensland Press, Brisbane, 1979.

Entrekin, Nicholas J., *The Betweeness of Place*, Johns Hopkins University Press, Baltimore, 1991.

Ferber, Sarah, Chris Healy & Chris McAuliffe (eds), *Beasts of Suburbia*, Melbourne University Press, Melbourne, 1994.

Fitzgerald, Shirley, *Sydney 1842–1992*, Hale & Iremonger, Sydney, 1992.

Fitzpatrick, David (ed.), *Home and Away: Immigrants in Colonial Australia*, Australian National University, Canberra, 1992.

Flanagan, Richard, 'Wilderness and History', *Public History Review*, I, 1992, pp. 103–17.

Flood, Josephine, *The Moth Hunters*, Australian Institute of Aboriginal Studies, Canberra, 1980.

—— 'Aboriginal Cultural Heritage of the Australian Alps: An Overview', in Babette Scougall (ed.), *Cultural Heritage of the Australian Alps*, Australian Alps Liaison Committee, Canberra, 1991, pp. 83–7.

Foster, Hal (ed.), *The Anti-aesthetic Essays on Post-modern Culture*, Bay Press, Port Townsend, 1983.

Frampton, Kenneth, 'Towards a Critical Regionalism: Six Points for an Architecture of Resistance', in Hal Foster (ed.), *The Anti-aesthetic. Essays on Post-modern Culture*, Bay Press, Port Townsend, 1983, pp. 16–30.

Franklin, Gordon, 'Let the Franklin Flow', sound recording, 1981.

Franklin, Miles, *My Brilliant Career*, Angus & Robertson, Sydney, 1901 (repr. 1966).

Frawley, K., 'The Gudgenby Property and Grazing in the National Park', *NPA Bulletin*, 25, 3 (March 1988), p. 4.

Freeland, J.M., 'People in Cities', in Amos Rapoport (ed.), *Australia as Human Setting*, Angus & Robertson, Sydney, 1972, pp. 99–123.

Fried, Marc, 'Grieving for a Lost Home', in L.J. Duhl (ed.), *The Urban Condition. People and Policy in the Metropolis*, Basic Books, New York, 1963, pp. 151–71.

Friends of Beecroft Reserve, *Newsletter*, 6 July, 1989.

Gale, F. & K. Anderson (eds), *Inventing Places*, Longman Cheshire, Melbourne, 1992.

Gans, Herbert,*The Urban Villagers*, The Free Press, New York, 1963.

—— *People, Places and Policies*, Columbia University Press, 1991.

Garnett, Julia, 'Bushland at Work', Friends of Beecroft Reserve, photocopy, n.d.

Garnett, Julia, 'The Turpentine Tree', ms.

Gilbert, Alan, 'The Roots of Anti-Suburbanism in Australia', in S.L. Goldberg & F.B. Smith (eds), *Australian Cultural History*, Cambridge University Press, Cambridge, 1988, pp. 33–49.

Goldberg, J., 'Sydney Motorways cannot be Justified', *Australian Financial Review*, 21 October 1993.

Goldberg, J., 'The F2 Castlereagh Expressway Affair: A Case for Reform of the NSW Roads and Traffic Authority', *Urban Policy and Research*, 11, 3 (1993), pp. 132–49.

Goldberg, S.L. & F.B. Smith (eds), *Australian Cultural History*, Cambridge University Press, Cambridge, 1988.

Garden, Don (ed.), *Created Landscapes*, History Institute of Victoria, Melbourne, 1992.

Goldsmith, Oliver, 'The Deserted Village', in Tom Davis (ed.), *Oliver Goldsmith, Poems and Plays*, Dent, London, 1975.

Goodman, David, 'Gold Fever/Golden Fields: The Language of Agrarianism and the Victorian Gold Rush', *Australian Historical Studies*, 29, 90 (April 1988), pp. 19–41.

—— 'The Politics of Horticulture', *Meanjin*, 47, 3 (1988), pp. 403–12.

Gray, Robert & G. Lehmann (eds), *Australian Poetry in the Twentieth Century*, Heinemann, Melbourne, 1994.

Griffiths, Tom, 'History and National History: Conservation Movements in Conflict', in J. Rickard and P. Spearritt (eds), *Packaging the Past*, Melbourne University Press, Melbourne, 1991, pp. 16–32.

Gunew, S. & Kateryna Longley (eds), *Striking Chords*, Allen & Unwin, Sydney, 1991.

Halket, J.P., *The Quarter Acre Block*, Australian Institute of Urban Studies, Adelaide, 1976.

Hall, Colin, *Wasteland to World Heritage*, Melbourne University Press, Melbourne, 1992.
Hawthorne, Lesleyanne (ed.), *Refugee, the Vietnamese Experience*, Oxford University Press, Melbourne, 1982.
Healey, Chris (ed.), *The Lifeblood of Footscray: Working Lives at the Angliss Meatworks*, Melbourne's Living Museum of the West, Footscray, c. 1982.
Hepworth, John, 'The Dancing Valley of the Mountain Horsemen', in Bryan Jameson (ed.), *Movement at the Station*, Collins, Sydney, 1987.
Heseltine, Harry (ed.), *A Tribute to David Campbell*, University of New South Wales Press, Sydney, 1987.
Hewett, Dorothy, *A Tremendous World in her Head*, Dangaroo Press, Sydney, 1989.
Hodges, Sue, 'A Sense of Place', in Don Garden (ed.), *Created Landscapes*, History Institute of Victoria, Melbourne, 1992, pp. 73–88.
Hodgins, Philip, *Down the Lake with Half a Chook*, ABC Enterprises, Sydney, 1988.
Hollensen, Anne, 'The Recollections of Jock Lawson', photocopy, Cranney Collection, Centre for Gippsland Studies, 1988.
Hosking, S., 'I 'ad to 'ave me Garden', *Meanjin*, 47, 3 (1988), pp. 439–53.
Howett, Catherine M., 'Notes Towards an Iconography of Regional Landscape Form: The Southern Model', *Landscape Journal*, 4, 2 (1985), pp. 75–85.
Hueneke, Klaus, *Huts of the High Country*, Australian National University Press, Canberra, 1982.
——*Kiandra to Kosciusko*, Tabletop Press, Canberra, 1987.
Huggins, Jackie & Rita Huggins, *Auntie Rita*, Aboriginal Studies Press, Canberra, 1994.
Isaacs, Eva, *Greek Children in Sydney*, Australian National University Press, Canberra, 1976
Jackson, Kenneth, *The Crabgrass Frontier*, Oxford University Press, New York, 1985.
Jackson-Nakano, Ann, The Death and Resurrection of the Ngunnawal: A Living History, M.Litt thesis, Australian National University, 1994.
Jameson, Bryan (ed.), *Movement at the Station*, Collins, Sydney, 1987.
Janovits, Anne, *England's Ruins. Poetic Purpose and the National Landscape*, Blackwell, London, 1990.
Johnson, Dick, *The Alps at the Crossroads*, Victorian National Parks Association, Melbourne, 1974.
Johnston, Ruth, *Immigrant Assimilation*, Peterson Brokensha, Perth, 1965.
——*The Assimilation Myth*, Publications of the Research Group for European Migration Problems XIV, Martinus Nijhoff, The Hague, 1969.

Jordan, R.D. & P. Pierce (eds), *The Poets' Discovery: Nineteenth Century Australia in Verse*, Melbourne University Press, Melbourne, 1990.

Jung, Karl, *Memories, Dreams, Reflections*, rec. and ed. Aniela Jaffe, tr. R. & C. Kingston, Collins Fontana, London, 1967.

Jupp, James (ed.), *The Australian People*, Angus & Robertson, Sydney, 1988.

Jurgensen, Manfred (ed.), *Ethnic Australia*, Phoenix, Brisbane, 1985.

Kamminga, Johann, 'Aboriginal Settlement and Prehistory of the Snowy Mountains', in Babette Scougall (ed.), *Cultural Heritage of the Australian Alps*, Australian Alps Liaison Committee, Canberra, 1991, pp. 101–24.

Kay, Jeanne, 'Landscapes of Women and Men: Rethinking the Regional Historical Geography of the United States and Canada', *Journal of Historical Geography*, 17, 4 (1991), pp. 435–52.

Keller, Stephen, 'Prelude to Power', ABC Radio documentary, c. 1959, held in Snowy Mountains Corporation Library, Cooma.

Ker Conway, Jill, *The Road to Coorain*, Heinemann, London, 1989.

Kinross Smith, Graeme & Jamie Grant, *Turn Left at any Time with Care*, University of Queensland Press, Brisbane, 1975.

Kroeber, Karl, *Romantic Landscape Vision, Constable and Wordsworth*, University of Wisconsin Press, Madison, 1975.

Krupinski, Jerszy & Graham Burrows, *The Price of Freedom: Young Indochinese Refugees in Australia*, Pergamon, Sydney, 1986.

Kunz, Philip, 'Immigrants and Socialisation: A New Look', *Sociological Review*, 16, 3 (November 1968): 363–75.

La Trobe Valley Express, 1 July 1970, 29 August 1974, 6 August, 20 August, 5 September 1975.

Lake Pedder Action Committees of Tasmania and Victoria, prod. and dir. by Peter Dodds & Ross Matthews, *The Struggle For Pedder*, film, 1971.

Lake Pedder Action Committees, *The Future of Lake Pedder*, June 1973, Lake Pedder Action Committees, Hobart, 1973.

Land Rights News, 2, 35 (April 1995).

Lawrence, D.H. & M.L. Skinner, *The Boy in the Bush*, Heinemann, London 1924 (repr. 1972).

Lawrence, D.H., *Kangaroo*, Heinemann, Melbourne, 1923 (repr. 1963).

Lawrie, Dawn, 'The Frustrations of a Civil Population Associated with a Major Reconstruction Project', photocopy, c. 1979.

Lawson, Henry, 'The Bush Undertaker', in Colin Roderick (ed.), *Henry Lawson. Short Stories and Sketches*, Angus & Robertson, Sydney, 1972, vol 1.

Lee, Claude, 'Burragorang Valley', in *A Place to Remember: Burragorang Valley 1957*, Bowral, 1971.

Leggett, Malcolm, Fake Pedder: Drowning or Enlarging a Lake, BA (Hons) thesis, Australian National University, 1994.

Lindbergh, Ann Morow, *Hour of Gold, Hour of Lead*, Harcourt Brace Jovanovich, New York, 1973.

Live Wire (Yallourn newspaper), various issues.

Llewellyn, Kate, *The Waterlily*, Hudson, Melbourne, 1987.

Lowenthal, David, 'The Place of the Past in the American Landscape', in D. Lowenthal & M.J., Bowden (eds), *Geographies of the Mind, Essays in Historical Geosophy*, Oxford University Press, New York, 1976.

Mackay, Marge, 'Adaminaby, the Old Town', ms.

Martin, Jean, *Refugee Settlers*, Australian National University Press, Canberra, 1965.

Martin, S., *A New Land*, Allen & Unwin, Sydney, 1993.

Marwood, Jim, *Valley People*, Kangaroo Press, Sydney, 1984.

Massey, Doreen, 'A Place Called Home', *New Formations*, 17 (Summer 1992), pp. 3–15

Matthews, Jill, *Good and Mad Women*, Allen & Unwin, Sydney, 1984.

McDonald, Nan, *Pacific Sea*, Angus & Robertson, Sydney, 1947.

McGoldrick, Prue, *Yallourn Was*, Gippsland Printers, Morwell, 1984.

McHugh, Siobhan, *The Snowy, the People Behind the Power*, Heinemann, Sydney, 1989.

Medical Journal of Australia, 24 May 1975, editorial.

Miller, Jim Wayne, *Brier, His Book*, Gnomon Press, 1983.

Milne, Gordon, 'Cyclone Tracy I. Some Consequences of the Evacuation for Adult Victims', *Australian Psychologist*, 12, 1 (March 1977), pp. 39–62.

Modjeska, Drusilla (ed.), *Inner Cities, Australian Women's Memory of Place*, Penguin, Melbourne, 1989.

Monk, Janice, 'Gender in the Landscape: Expressions of Power and Meaning', in F. Gale & K. Anderson (eds), *Inventing Places*, Longman Cheshire, Melbourne, 1992, pp. 122–138.

Mulvaney, D.J., 'The Alpine Heritage in Perspective', in Babette Scougall (ed.), *Cultural Heritage of the Australian Alps*, Australian Alps Liaison Committee, Canberra, 1991, pp. 9–17.

—— (ed.), *The Humanities and the Australian Environment*, Australian Academy of the Humanities, Canberra, 1991.

Nash, Roderick, *Wilderness and the American Mind*, Yale University Press, New Haven.

National Capital Development Commission, 'The Gudgenby Area. Policy Plan and Development Plan', National Capital Development Commission, Canberra, September 1984.

National Parks & Wildlife Service, 'Proposed Gudgenby National Park Land Use Study', Australian National Parks and Wildlife Service, Canberra, 1976.

Nicolson, Marjorie, *Mountain Gloom and Mountain Glory*, Norton, New Haven, 1967.

Northern Territory News, Special Supplement, 'Cyclone Tracy', c. 12 December 1994.

Northern Territory News, 'Cyclone Tracy: A Story of Survival', c. 1 December 1994.

O' Farrell, Patrick, 'Defining Place and Home. Are the Irish Prisoners of Place?', in David Fitzpatrick (ed.), *Home And Away: Immigrants in Colonial Australia*, Australian National University, Canberra, 1992, pp. 1–18.

Papaellinas, George (ed.), *Homeland*, Allen & Unwin, Sydney, 1991.

Parker, Gordon, 'Psychological Disturbance in Darwin Evacuees Following Cyclone Tracy', *Medical Journal of Australia*, 21 (24 May 1975), pp. 650–2.

Patterson, Rex, Minister for the Department of Northern Territory, Press release, 1 April 1975.

Pearce, Owen W., *Rabbit Hot, Rabbit Cold. Chronicles of a Vanishing Australian Community*, Woden, 1991.

Pearson, Michael, Seen Through Different Eyes, PhD thesis, Australian National University, 1981.

Pitt, Marie J., *Selected Poems of Marie J. Pitt*, Lothian, Melbourne, 1944.

Pizer, Marjorie, *Selected Poems 1963–1983*, Pinchgut Press, Sydney, 1984.

Porteous, Douglas J., 'Home the Territorial Core', *Geographical Review*, LXVI (1976), pp. 383–90.

Praed, Rosa, *Lady Bridget in the Never Never*, Pandora, London, 1915 (repr. 1987).

Price, Charles (ed.), *Australian Immigration*, Australian National University, Canberra, 1966.

Prichard, Katharine Susannah, *Coonardoo*, Angus & Robertson, Sydney, 1927 (repr. 1975).

Proudfoot, Helen, *Gardens in Bloom*, Kangaroo Press, Sydney, 1989.

Public Works Committee Victoria, 'Yallourn Coal Reserves Inquiry', transcript of evidence given before the Public Works Committee at Yallourn.

Pybus, Cassandra (ed.), *Columbus' Blindness*, University of Queensland Press, Brisbane, 1994.

Ramson, William S., 'Wasteland to Wilderness: Changing Perceptions of the European Environment', in D.J. Mulvaney (ed.), *The Humanities and the Australian Environment*, Australian Academy of the Humanities, Canberra, 1991, pp. 5–19.

Rapoport, Amos (ed.), *Australia as Human Setting*, Angus & Robertson, Sydney, 1972.

Raymond, Robert, *A Town to be Drowned*, film, copy in Snowy Mountains Corporation library, 1957.

Read, Peter, 'Being is a Transitive Verb: Four Views of a Life Site in Mittagong New South Wales', *Landscape Research*, 19, 2 (Summer 1994), pp. 58–67.

—— 'Joy and Forgiveness in a Haunted Country', *New Norcia Studies*, 1994, pp. 1–9.

—— 'Mystery that Seeps from the Earth', forthcoming.

—— 'Our Lost Drowned Town in the Valley', *Public History Review*, 1 (1992), pp. 160–73.

Relph, Edward, 'Geographical Experiences and Being-in-the World: The Phenomenological Origins of Geography', in David Seamon & Robert Mugerauer (eds), *Dwelling, Place and Environment*, Martinus Nijhoff, Dordrecht, 1985, pp. 15–31.

Report of the Committee of Inquiry into the National Estate, Australian Government Publishing Service, Canberra, 1974.

Report of the Royal Commission into the Place of Origin and the Causes of the Fires which Commenced at Yallourn ..., Victorian Government Printer, 1944.

Richards, Lyn, *Nobody's Home. Dreams and Realities in a New Suburb*, Oxford University Press, Melbourne, 1990.

Rickard, J. & P. Spearritt (eds), *Packaging the Past*, Melbourne University Press, Melbourne, 1991.

Ricoeur, Paul, 'Civilization and National Cultures', in P. Ricoeur, *History and Truth*, Northwestern University Press, Evanston, 1965, pp. 271–84.

Riemer, Andrew, *Inside Outside*, Angus & Robertson, Sydney, 1992.

—— *The Habsburg Cafe*, Angus & Robertson, Sydney, 1993.

Roads and Traffic Authority NSW, *Community Newsletter No. 2*, December 1993.

—— *F2-Castlereagh Freeway, Environmental Impact Statement*, New South Wales Government Printer, Sydney, 1989.

Robins, Kevin, 'Reimagined Communities? European Image Spaces, Beyond Fordism', *Cultural Studies*, 3, 2 (May 1989), pp. 145–65.

Roderick, Colin (ed.), *Henry Lawson. Short Stories and Sketches*, Angus & Robertson, Sydney, 1972.

Rose, Deborah B., *Dingo Makes Us Human*, Cambridge University Press, Cambridge, 1992.

—— 'Review of K. Neumann, *Not the Way it Really Was*', *Australian Historical Studies*, 103 (October 1994), pp. 310–11.

Rose, Gillian, *Feminism and Geography*, University of Minnesota Press, Minneapolis, 1993.

Rosh White, Naomi, *From Darkness to Light*, Collins Dove, Melbourne, 1988.

Salvesen, Christopher, *The Landscape of Memory*, Edward Arnold, London, 1965.

Saunders, Peter, 'The Meaning of "Home" in Contemporary English Culture', *Housing Studies* 4, 3 (July 1989), pp. 177–92.

Save Yallourn Committee, *To Yallourn With Love*, Mitchell River Press, Bairnsdale, 1984.

Scougall, Babette (ed.), *Cultural Heritage of the Australian Alps*, Australian Alps Liaison Committee, Canberra, 1991.

Seddon, G. & Mari Davis (eds), *Man and Landscape in Australia, Towards an Ecological Vision*, Australian Government Publishing Service, Canberra, 1976.

Shaw, Tom, 'Natural Comedians', short story, typescript, Centre for Gippsland Studies.

Sherington, Geoffrey, *Australia's Immigrants*, Allen & Unwin, Sydney, 1980.

'Slab' [Ted Hopkins], *The Yallourn Stories*, Champion Books, Melbourne, c.1982.

Smith, Bernard, 'On Perceiving the Australian Suburb', in G. Seddon & Mari Davis (eds), *Man and Landscape in Australia, Towards an Ecological Vision*, Australian Government Publishing Service, Canberra, 1976, pp. 289–304.

Snowy Mountains Authority [D.H. White], *Operation Adaminaby*, film, held at Snowy Mountains Corporation library, 1958.

——— *The Snowy Mountains Story*, Snowy Mountains Authority, Cooma, n.d., 1970s.

Soja, Edward M., *Postmodern Geographies: The Reassertion of Space in Contemporary Social Theory*, Verso, London, 1989.

Speller, Gerda M., Landscape, Place and the Psycho-social Impact of the Channel Tunnel Terminal Project, MSc thesis, University of Surrey, 1988.

State Electricity Authority Victoria [Mary Wilton], *A Town Born to Die*, film and transcript of soundtrack, Centre for Gippsland Studies.

Stewart, Douglas, *Collected Poems 1936–1967*, Angus & Robertson, Sydney, 1967.

Stewart, Nancy, *Yarns, Stories of Creative Women in Melbourne's West*, Melbourne's Living Museum of the West, Footscray, 1985.

Strehlow, T.G.H., *Aranda Traditions*, Melbourne University Press, Melbourne, 1947.

Stretton, Alan, *The Furious Days*, Collins, Sydney, 1976.

Sydney Morning Herald, 'Namadji, Canberra's Highland Wilderness', 24 June 1993.

Tacey, David, 'Australian Landscape as a Spiritual Problem', *Canberra Jung Society Newsletter*, February–July 1993, pp. 13–17.

Tasfilm, *Lake Pedder*, 1971.

Taylor, Ken, 'Cultural Values in Natural Areas', in Babette Scougall (ed.), *Cultural Heritage of the Australian Alps*, Australian Alps Liaison Committee, Canberra, 1991, pp. 55–65

——— 'Defining an Australian Sense of Place: Cultural Identity in Landscape and Painting', in *CELA 94*, History and Culture Conference Proceedings [Council of Educators in Landscape Architecture], pp. 270–80.

Terkel, Studs, *Division Street: America*, Allen Lane, Penguin Press, London, 1968.

The New Darwin, 1, 1 (26 August 1975).

Thompson, Stephanie, *Australia Through Italian Eyes*, Oxford University Press, Melbourne, 1980.

Tighe, C., 'The Origins of Namadgi National Park', *National Parks Association Bulletin*, 29, 1 (March 1992), pp. 14–21.

Todd, Ken, 'Darwin Post-Tracy', *Royal Australian Planning Institute Journal*, 17 (August 1979), pp. 192–6.

Trades and Labour Council, Central Gippsland, Report to VTLC on the Yallourn Township, 9 May 1974, Centre for Gippsland Studies.

Tuan, Yi-Fu, *Space and Place*, Edward Arnold, London, 1977.

Unger, Margaret, *Voices from the Snowy*, University of New South Wales Press, Sydney, 1989.

Urban Action, 'Careerism, Incompetence or Funny Finance', Special Supplement, 8 (Winter 1993), pp. 1–2.

Viviani, Nancy, *The Vietnamese in Australia: New Problems in Old Forms*, Griffith University, Nathan, 1980.

Vleeskens, Cornelius, 'This Eternal Curiosity: The Search for a Voice in the Wilderness', in S. Gunew & Kateryna Longley (eds), *Striking Chords*, Allen & Unwin, Sydney, 1991.

Wardley, David & Michael Ballock, 'Satisfaction and Positive Resettlement: Evidence from Yallourn, Latrobe Valley, Australia', *American Planning Association Journal*, 46, 1 (Jan. 1980), pp. 64–75.

Waten, Judah, *Alien Son*, Angus & Robertson, Sydney, 1952.

Webber, D., 'Darwin Cyclone: An Exploration of Disaster Behaviour', *Australian Journal of Social Issues*, 11, 1 (1976), pp. 54–63.

Weekend Australian, 'Cyclone Tracy', 17 December 1994.

Weekend Australian, Special Supplement, 'Cyclone Tracy Twenty Years On', 24 December 1994.

White, D.H., draft and amendments, 'The Town of Adaminaby and the Snowy Mountains Scheme', typescript, Snowy Mountains Corporation Library, Cooma.

Williams, Raymond, *The Country and the City*, Oxford University Press, Oxford, 1973.

Woodward, John, Commissioners of Inquiry for Environment and Planning, Report to the Honourable David Hay, ... *A Proposed Expressway from Pennant Hills Road, Beecroft to Pittwater Road, Ryde, known as the F2 Stage 1*, July 1990, New South Wales Government Printer, 1990.

Wright, Judith, *A Human Pattern*, Angus & Robertson, Sydney, 1990.

Wynnhausen, Elizabeth, *Manly Girls*, Penguin, Melbourne, 1989.

Zable, Arnold, *Jewels and Ashes*, Scribe, Newham, 1991.

—— 'Voices from the Silence', in Cassandra Pybus (ed.), *Columbus' Blindness*, University of Queensland Press, Brisbane, 1994, pp. 30–40.

INDEX

Printed in the United States
By Bookmasters